1994

THE DEVIL WE KNEW

THE DEVIL
WE KNEW

Americans and the Cold War

H. W. BRANDS

New York   Oxford
OXFORD UNIVERSITY PRESS
1993

Oxford University Press

Oxford   New York   Toronto
Delhi   Bombay   Calcutta   Madras   Karachi
Kuala Lumpur   Singapore   Hong Kong   Tokyo
Nairobi   Dar es Salaam   Cape Town
Melbourne   Auckland

and associated companies in
Berlin   Ibadan

Published by Oxford University Press, Inc.,
200 Madison Avenue, New York, New York 10016

Oxford is a registered trademark of Oxford University Press

Library of Congress Cataloging-in-Publication Data
Brands, H. W.
The devil we knew : Americans and the Cold War / H.W. Brands.
p.   cm.
ISBN 0-19-507499-8
1. United States—Foreign relations—1945–1989.
2. United State -Foreign relations—1989-
3. Cold War.   I. Title.
E744.B6974   1993   327.73—dc20
92-29030

2 4 6 8 9 7 5 3 1

Printed in the United States of America
on acid-free paper

Preface

Individuals are known by the company they keep, nations by the company they avoid. Historically, ties among nations have rarely reflected common hopes so much as common fears. Indeed, the very existence of most nations owes to a simultaneous perception of external danger on the part of previously disunited persons and groups, who then band together to form the nations. For building national solidarity, nothing beats enemies at the gate.

Americans are no more exceptions to this rule than they are to other norms of human behavior. Arguably, they are less exceptional. As a nation chiefly of immigrants, America has lacked many of the usual characteristics of nationhood, especially a shared ancestry, language, and cultural experience. For this reason, Americans' shared enemies have mattered the more. The American Declaration of Independence announced adherence to the positive principles of life, liberty, and the pursuit of happiness, but the document derived its bite—and the American revolt against Britain its energy—from the crimes Jefferson charged to King George's account. For the first century of America's national existence, the British continued to play bogey. As late as 1895, the customarily mild Grover Cleveland could expect to revive his failing fortunes by twisting the lion's tail regarding an obscure dispute in the wilds of South America. The jingoes loved it, though not quite enough to save the Democratic party from William Jennings Bryan.

After the beginning of the twentieth century, Britain made a less convincing villain than before, primarily because London went out of its way to be nice. But new villains soon appeared—which largely explained the Brits' sudden sensitivity to American concerns. Germans, alias Huns and Belgian-baby butchers, served as foils to America's innocence during World War I. The Germans returned for an encore two decades later, sharing billing this time with the Japanese.

The most reliable black hats since 1917, however, were the communists of the Soviet Union, and their ideological kin in other countries. The first

152, 137

Red Scare of the late 1910s and early 1920s, which was accompanied by an American invasion of the revolutionary Bolshevik state, was followed by a decade and a half of attempts to quarantine the communist infection. From the mid-1930s through the mid-1940s, in the face of the greater fascist threat, America dropped its guard temporarily. But as soon as Hitler was dispatched, Americans once more heard Marx rattling the chains of the proletariat, and again saw Lenin purchasing rope with which to hang the capitalists. After 1945, fear of communism became institutionalized in the United States. A new Red Scare coincided with the onset of the Cold War, and helped inspire efforts to isolate the second focus of infection, China. The consequences were as unsuccessful as the attempts to isolate Russia had been. The failure didn't result from lack of trying. For two generations, until the end of the 1980s, Americans girded to fight communism on land, on sea, and in the air, in the arctic and tropics, from the barren hills of Greece to the jungles of Southeast Asia. In two instances, American opposition produced full-blown wars, killing more than 100,000 Americans and millions of un-Americans. Although the 1970s briefly moderated the Bolshie-bashing, in the age of Ronald Reagan the communist specter once more haunted America.

For all its psychological implications, the Cold War was, quite obviously, more than the latest episode in an ongoing search for enemies by which to affirm America's identity and Americans' basic goodness. It was also a strategic struggle. This aspect of the Cold War was most evident to those who fashioned American policy during the Cold War's early years. American leaders saw the Soviet Union as a serious danger to American national security. Postwar Europe lay in ruins, while the Soviet army occupied half the continent and threatened the rest. Communist parties contested for power in countries the Kremlin didn't control. Perhaps these parties took orders from Moscow, perhaps not. But in either case, they had more in common with the socialist East than with the capitalist West, and their victory would aggravate democracy's danger. The United States, out of a simple instinct for survival, had to take measures to offset Soviet strategic weight and circumscribe Soviet influence.

Beyond matters of psychology and strategy, the Cold War was an economic struggle. This facet of the superpower rivalry seemed clearest to historians and political scientists writing during the 1960s and shortly thereafter, when the costs of containing communism were mounting dramatically. Viewed in economic terms, at least by the revisionists, the Cold War had resulted largely from the efforts of the United States to export capitalism across the globe. American leaders, concerned that a repetition of the depression of the 1930s would trigger the collapse of the American way of life, and convinced that preventing a repetition required opening foreign markets to American products, sallied forth to bring as much of the world as possible into the American economic sphere. The Soviet Union, far from being the aggressor, found itself on the defensive. Although at times the Kremlin reacted belligerently, the burden of responsibility for the Cold War rested on the United States, the more powerful of the two countries (possessing, among other advantages, a nuclear monopoly), and the one with less to fear from the other.

Finally, the Cold War was a long-running issue in domestic American politics. American politicians could make reputations out of the communist threat, as the rise of Joseph McCarthy, a man with little else to recommend him, amply showed. Although the red-baiting subsequently grew more subtle, it remained a staple of American political discourse, and persons opposed to all manner of changes in the American status quo regularly trotted out the subversive charge to discredit reformers. The old nag eventually died, and Americans' obsession with conspiracies faded. But, as Reagan demonstrated, the constituency for fervent denunciations of communism remained strong well into the 1980s.

At times past, various authors writing on the Cold War have tried to isolate one cause or another. Traditionalists stressed the Soviet strategic threat, while revisionists concentrated on economic elements in the superpower struggle. If history were a science, such reductionism might make sense. But it's not, and it doesn't. Historical reductionism makes especially little sense for the United States, a huge country comprising hundreds of millions of people with a daunting diversity of needs, desires, hopes, and fears. America almost never speaks with a single voice, and even when it does, this voice never reflects a single line of thinking or motivation. Individuals do what they do from a variety of motives. All the more do great agglomerations of individuals.

The fundamental question addressed in this book is: Why did the United States act as it did in the Cold War? The answer is: For a variety of reasons. Sometimes the reasons mostly involved the condition of the American psyche, and the manner in which Americans defined themselves vis-à-vis the world. Sometimes the reasons clustered around strategic concerns, notably the defense of the United States and American institutions against a heavily armed and avowedly hostile foe. Sometimes the reasons turned on economic matters, on the the real and perceived needs of the American economy collectively and of certain sectors of the economy separately. Sometimes the reasons were primarily political, having to do with getting candidates into office, keeping them there, and enacting their personal and party programs.

One aspect of the fundamental question bears reiterating. The present study attempts to explain why *Americans* acted as they did during the Cold War. Every international conflict, by definition, requires at least two national antagonists. The Cold War had two major antagonists, a handful of seconds, and auxiliary forces of several more on each side. Why the Soviets did what they did, and what got into the British, French, Chinese, Koreans, Vietnamese, Cubans, and the rest is for someone else to say.

If the purpose of this book were to assign blame for the trials the Cold

War visited upon humanity, then concentrating on the United States would skew the case. No country is heroic to its historians, or at least no country ought to be. While the analysis here reflects skepticism regarding the uniform high-mindedness of American leaders and the American people, the reader might (or might not) be interested to know that the author suspects that close treatment of other countries' policies would produce national profiles at least as unflattering. It's a thoroughly imperfect world we inhabit.

A second point also bears keeping in mind: namely, the difference between the factors that *precipitated* the Cold War and those that *perpetuated* it. Americans adopted a policy of militant opposition to the Soviet Union and communism during the late 1940s for one set of reasons. They maintained this policy for most of four decades for another set of reasons. The two sets overlapped, to be sure. But the Cold War, once established and institutionalized on the American side (as, no doubt, on the Soviet side), developed a life of its own. The Cold War and everything it represented—psychologically, strategically, economically, politically—became terra cognita for the major groups shaping American policy. They were reluctant to leave it.

The volume at hand does not pretend to comprehensiveness. It should be read as an essay rather than a history. Certain developments receive more attention than they would in a full-scale treatment of American participation in the Cold War. Others receive less, even none at all. While enough information about the basic facts of the Cold War has been included to make the arguments understandable to readers with no special expertise, the emphasis is on the meaning beneath the facts.

It will be the rare—but, needless to say, incisive—reader who agrees with all the interpretations offered here. Even rarer—and rather less incisive—will be the reader who finds them entirely novel. But neither agreement nor novelty is the object of the exercise. The object is to provoke thought about the period that, though evidently over now, produced the world we are grappling with today.

Austin, Texas   H.W.B.
February 1993

Contents

THE DEVIL WE KNEW

CHAPTER ONE

★

The Last Days of
American Internationalism
1945–1950

The spirit of Yalta

The Cold War had a double taproot. One branch ran to 1783, when the United States won recognition of its independence from Britain and acquired the original American West. Independence made the United States a player on the stage of international affairs, while the trans-Appalachian transaction set Americans on a course of expansion that would characterize the first two centuries of the republic's existence. During the initial century and a quarter, the United States expanded mostly westward, but after 1914 Americans devoted increasing energy to the affairs of Europe. Meanwhile, Russia was expanding similarly, although Moscow reversed the compass directions. Moscow's nineteenth-century rulers looked to the east, to the same Pacific the Americans were reaching for, and only after the second German invasion of the twentieth century did the Kremlin push its sphere west into the heart of Europe. The two expanding empires had bumped briefly in the 1860s, when the United States dispossessed Russia of Alaska, but the circle didn't close until 1945, when the American and Russian armies met at the Elbe in a Europe suddenly emptied of credible competition.

If the geopolitical origins of the Cold War lay in the eighteenth and nineteenth centuries, the ideological aspect of the conflict sprang only from the second decade of the twentieth. In 1917, Woodrow Wilson launched a campaign to make the world safe for democracy, while in the same year Lenin led a revolution to make the world safe for socialism. In theory, the objectives of the two men might not have been incompatible. But neither Wilson nor Lenin could conceive of democratic socialism. Wilson's democracy implied capitalism—economic democracy—while Lenin's socialism rested on dictatorship by the communists, the vanguard

3

of the proletariat. A third ideology—fascism—kept the democrats and the communists from having at each other until 1945, even as the geopolitics of the era kept the American and Russian empires apart. But the collapse of fascism consequent to the defeat of Germany and Japan left the Americans and Russians by themselves to vie for the ideological allegiance of the planet's peoples.

The dual character of the Cold War—being both a geopolitical and an ideological contest—rendered the simultaneous disappearance of geopolitical competition and of challenging ideologies doubly important in precipitating it. As geopolitical great powers, the United States and the Soviet Union couldn't avoid looking on each other with the suspicions countries have always had for other countries capable of doing them serious harm. To each, the other was a strategic threat, and what augmented the military, economic, or political power of one necessarily jeopardized the physical security of the other. As the principal proponents of effectively contradictory ideologies, the United States and the Soviet Union couldn't avoid viewing each other with the suspicions true believers have always had for infidels and heretics. To each, the other was an existential threat, and what augmented the philosophical or moral influence of one necessarily jeopardized the psychological security of the other.

The dual character of the conflict between the United States and the Soviet Union became apparent even before World War II ended and thereby cleared the clutter between the two countries and the causes they represented. In February 1945, Franklin Roosevelt met Joseph Stalin at Yalta to discuss matters of mutual concern. Britain's prime minister Winston Churchill joined the American president and the Soviet premier.

Four issues engaged the threesome. The first, disposed of with little difficulty, produced a Soviet promise to enter the Pacific war within three months after the defeat of Germany. In exchange, Moscow would receive the Kurile Islands and guarantees of a return to the status quo in northeastern Asia as it had existed prior to the Russo-Japanese War of 1904–5.

The second issue involved the future of Germany. The three leaders ratified an earlier decision-in-principle to divide Germany. They specified one zone of occupation each for the United States, Britain, and the Soviet Union, and they allowed that France might also claim a zone. Whether this initial partition would lead to permanent dismemberment, they did not definitely state. Stalin wanted such a statement, determined that Germany not survive the war intact. Roosevelt didn't dismiss dismemberment, but in the end the three leaders concurred only on measures to destroy Naziism and German militarism, and "such other measures in Germany as may be necessary to the future peace and safety of the world."[1]

The third issue had to do with the nature and purpose of the postwar international security organization. Stalin showed little interest in this bourgeois idea. According to the American minutes of a February 4 meeting, "Marshal Stalin made it quite plain on a number of occasions that he felt that the three Great Powers which had borne the brunt of the war and had liberated from German domination the small powers should have the unanimous right to preserve the peace of the world." The Soviet premier dismissed as ridiculous the idea of consulting the lesser countries on weighty issues of global affairs. "Marshal Stalin said that he was prepared in concert with the United States and Great Britain to protect the rights of the small powers, but that he would never agree to having any action of the Great Powers submitted to the judgment of the small powers." In short, the world organization might be a nice place for talk, but for real security the great powers must look to themselves—as great powers always had.[2]

The American position on what would become the United Nations organization was rather different. Roosevelt personally had few problems with the great powers policing the world. The president told Stalin that indeed the three victors should write the peace treaty and assume the "greater responsibility" for maintaining the peace. Yet Roosevelt recognized that permanent policing might not sit well with the American people. While he said he thought Congress would go along with reasonable measures to safeguard the future peace, he didn't think such support would extend to a continuing American military presence in Europe. He predicted that Americans troops might remain in Europe for two years, but not much more. (At this expression of concern for the opinion of the American people, Stalin's aide Andrei Vyshinski told Roosevelt's interpreter Charles Bohlen that the American people should learn to obey their leaders. Bohlen responded that he would like to see Mr. Vyshinski undertake to teach this lesson to the American people. Vyshinski said he would be delighted to, given the opportunity—which he later was, when he became Stalin's representative at the United Nations. The lesson wasn't well learned, and the teacher became one of America's favorite ogres.)[3]

Political constraints, if nothing else, predisposed American officials to place more store than the Soviets in collective security, by which the Americans meant the preservation of peace through the voluntary collaboration of the natural majority of law-abiding countries. This had been the aim of the League of Nations, and while the League's failure suggested to the Soviets a fatal flaw in the very concept of collective security, Americans tended to interpret the failure as a matter of poor execution. American leaders thought the idea of the League deserved another try, although they hedged by insisting that the great powers possess a veto in the new international organization's inner council, which would handle the most important matters.

The fourth Yalta issue—the real sticker—was the Polish question. Here the divergence between the Soviet and American views became clearest. In one conversation, Churchill noted that the future of Poland touched Britain's honor, in that Britain had entered the war in response to Hitler's invasion of Poland. Stalin replied that Poland for the Soviet Union was more than a matter of honor: "It is also a question of the security of the state, not only because we are on Poland's frontier but also because throughout history Poland has always been a corridor for assaults on Russia." In the last thirty years alone, Poland had twice served as a staging area for German attacks on Russia. The future condition of Poland, Stalin said, was a matter "of life and death for the Soviet state." Stalin went on to assert that the best guarantee of Soviet security in this area was a "free, independent and powerful" Poland. What he meant was that Poland should be free and independent of influences hostile to the Soviet Union, and powerful enough to repel any further attacks from the west. To this end, he advocated turning control of the country over to a provisional regime established in Lublin—a group carefully vetted and guided by Moscow.

Roosevelt, backed by Churchill, preferred the Polish government-in-exile in London. This government offered a greater measure of Polish self-determination than the Lublin regime, and, because it did, it was much more popular with those "six or seven million Poles in the United States" whom, the president reminded Stalin, the American government could not ignore. Roosevelt told Stalin he appreciated Russian concerns. "We want a Poland that will be thoroughly friendly to the Soviet [sic] for years to come. This is essential." But he said he didn't believe that Polish friendliness precluded involvement by the London Poles in a government for liberated Poland. He advocated bringing the London Poles and the Lublin Poles together to form "a government of national unity."[4]

Roosevelt judged the difference between the United States and the Soviet Union on Poland to be not unbridgeable, and he did his best to bridge it. He told Stalin, "I hope I do not have to assure you that the United States will never lend its support in any way to any provisional government in Poland that would be inimical to your interests." Roosevelt went on to push for open elections in Poland at the earliest possible date. He concluded, "I know this is completely consistent with your desire to see a new free and democratic Poland emerge from the welter of this war."[5]

Although Stalin agreed to let the London Poles enter Poland and join the provisional government, and despite his promise in the final Yalta communiqué to allow "free and unfettered elections as soon as possible on the basis of universal suffrage and secret ballot," the difference between Washington and Moscow on Poland was in fact beyond bridging. The United States wanted a democratic Poland, while the Soviets wanted a friendly Poland. Roosevelt might contend that these two objectives were reconcilable, but contending didn't make them so. The Poles couldn't forget their generations of ill treatment at Russian hands, nor would they forget that Stalin had been scarcely slower than Hitler to invade Poland in 1939. Almost of historical necessity, a genuinely democratic Poland wouldn't be friendly to the Soviet Union—certainly not as Stalin defined friendly.[6]

Roosevelt may, in truth, have recognized this fact, but the president chose not to push the Polish issue to the point of confrontation. Perhaps he thought he could finesse it in the coming months. Perhaps he wanted to see how Poland played at home once the other complications of the post-war period pushed their way into the American political arena. Perhaps the Poles would realize the precariousness of their position and arrange some modus vivendi with Moscow. In any case, Roosevelt didn't want the wartime alliance to fall apart publicly before the war had even ended. Besides, with the Soviet army occupying nearly all of Poland, he realized he was in no position to demand that Stalin accept the London Poles and allow elections. He must appeal to Stalin's better nature, such as it was. Roosevelt understood that on this issue the United States lacked the lever-age to enforce American preferences. Lastly, though Germany would surely surrender soon, Japan had plenty of fight left in it, and Soviet participation in the end game in the Pacific would go far toward keeping American casualties down in the expectedly expensive invasion of the Japanese home islands.

Events of the months after the Yalta conference overturned various of the calculations that informed American policy there. Roosevelt died in April, leaving direction of American diplomacy to Harry Truman. Tru-man's gifts didn't include subtlety, and with his assumption of office the possibility of a finessed solution to the Polish question vanished. The Soviets allowed the return of the London Poles, but procrastinated on the matter of elections. Ultimately, amid hardening attitudes in both East and West, the Kremlin canceled the elections entirely. Meanwhile, American and British scientists completed work on the atomic bomb. When dropped twice on Japan, the bomb brought the war in the Pacific to a far swifter conclusion than anyone had counted on at Yalta, thus rendering unneces-sary, after the fact, Roosevelt's solicitude for Soviet good will.

The most important consequence of Yalta turned out to be the division of Germany. When relations between the big powers soured, the zones of occupation, including the proffered one the French readily availed them-selves of, provided the basis for the semi-permanent division of Germany into two separate sovereign states. The division of Germany in turn pro-vided the model and much of the rationale for the division of the rest of Europe into Eastern and Western zones.

As a result, Yalta came to symbolize the Cold War order on the conti-nent. American critics of the new reality attacked Roosevelt's actions at Yalta as delivering millions more souls to communist domination. The criticism was greatly overblown. For the most part, Yalta simply ratified what the United States couldn't have changed without running unreason-able risks. Few of those who criticized Roosevelt's actions at the confer-ence held that he could have undone Moscow's military hammerlock on Poland or the other countries being wrested from Germany by Soviet troops. Supporters of Chiang Kai-shek's Chinese Nationalist regime stood on slightly firmer ground in denouncing the dealing that restored the Russian position in Manchuria, since Chiang, who in 1945 headed the closest thing China had to a legitimate government, had no voice at Yalta. Yet even the relevant agreement stated that China would retain "full sovereignty" in Manchuria, and it specified that the Russian restoration would take place only with Chiang's concurrence.

In symbolizing the division of Europe into Soviet and American spheres of influence, Yalta also represented the premature demise of any inclusive concept of collective security. The original idea was that the normal majority of peaceful and orderly countries would join forces in deterring or punishing law-breaking by the occasional deviant country, just as the great majorities of peaceful and orderly citizens in single countries join in deterring or punishing the occasional deviant individuals. This idea required general agreement among the great powers on issues of common concern. Yalta and subsequent events demonstrated that such agreement didn't exist. American leaders would attempt to salvage a severely circumscribed version of collective security, on which they would erect a system of alliances against the Soviet Union. But rather than being a formula for peace, as the original inclusive concept was, the modified, exclusive version would be a formula for conflict.

Remembrance of things past

An American worker who turned 65 at the end of World War II could have looked back on participation in America's labor force for five decades, give or take a few years depending on class and occupation. If he—the great majority of those working for pay over such a long period were males—had left school in the early 1890s, he would then have encountered considerable difficulty finding employment. In 1893, tremors on Wall Street triggered a financial panic that sent the country into its worst depression to date. For much of the rest of the decade, Americans suffered through unprecedented unemployment accompanied by massive social upheaval. The depression and the upheaval spawned demands for overseas expansion as a means to obtain outlets for American production and, less explicitly, to vent the energy that was ripping American society apart. The expansionist urges culminated in the war with Spain, which confirmed American hegemony in the Western Hemisphere, provided America an Asian colony (the Philippines), and set America squarely on the path to global power. For the most part coincidentally, the last years of the 1890s also witnessed an end to the depression. But tracing cause and effect in either economics or politics has never been an exact science, and many Americans could be forgiven for remembering that the same William McKinley who campaigned for and produced full dinner-pails also whipped the Spaniards and annexed the Philippines. Our typical worker was happy to be regularly employed for the first time.

He kept his job through the beginning of the new century, although he unknowingly suffered a close call in 1907 when Theodore Roosevelt had to summon J. P. Morgan to prevent another meltdown of American financial markets. When the next slump did come, toward the end of the Taft administration, it was nowhere near as devastating as the depression of the 1890s, but it sufficed to remind Americans that prosperity was hardly a necessary consequence of the unfettered operation of the American capital- ist system. If it hadn't sufficed, the loud complaints of the progressives about how monopolists and other malefactors of great wealth were squeez- ing the lifeblood out of the country, and how this necessitated an active role for government in evening the odds, would have.

The American economy was still sliding downward in August 1914, when war broke out in Europe. Commodity prices immediately soared, reviving the chronically weak farm sector. War orders for manufactured goods soon injected similar new life into industry. Unwilling to risk ruin- ing the recovery, Wilson overruled Secretary of State William Jennings Bryan, that inveterate enemy of big business and big-city finance, and allowed American banks to float belligerent loans in the United States. Bryan warned that this decision would lead to American involvement in the fighting, and he was right. The vast majority of the loans went to the British and French, who enjoyed naval superiority and hence bought more American goods than the Germans, and who had closer links to American financiers. Whatever its strategic implications, the loan decision kept the economic revival going. The war years were the most prosperous in a generation. The experience re-emphasized a lesson Northerners had learned during the Civil War: that there is nothing like a war fought elsewhere for making the economy boom.

Perhaps our worker weathered the postwar shakeout with job intact; perhaps not. If he tucked away a nest egg during the flush years of the 1920s, he might have invested it in the stock market. More conservatively, he may simply have banked it. If stocks were his choice, he almost cer- tainly lost his savings in the crash of 1929. If he put his future in the hands of his local banker, the banker quite possibly lost it in the crash or in the epidemic of defaults that followed. In either case, or even if he was one of the lucky persons who experienced the onset of the Great Depression of the 1930s only through the misfortunes of relatives and friends, his already shaky faith in the lasting compatibility of peace and prosperity would have been weakened further.

This faith probably vanished altogether during the decade of despair that separated the stock crash from the beginning of World War II. Most Americans didn't join the Communist party, although record numbers did. Most preferred the moderation and responsibility of Franklin Roose- velt to the pie-in-the-sky preachments of Huey Long and Francis Townsend and the right-wing rantings of radio priest Father Coughlin. But as unemployment stuck above 10 million and Roosevelt's best efforts failed to generate recovery, few Americans could feel hopeful about the ability of the American economy to provide them a stable, prosperous existence.

Once again, war came to the rescue. The recovery began before Pearl Harbor, as Washington built up the American army and navy and shipped weapons and other items to Britain, China, and the Soviet Union. It kicked into high gear when the United States entered the conflict. With farmers planting fencerow to fencerow, and factories running second and third shifts, all major measures of economic activity—employment, industrial and agricultural production, personal and national income—shot to levels never dreamed of before. Regardless of the fighting's effect on the fate of suffering humanity in Europe and Asia, it lifted America out of the Great Depression and restored prosperity to the land.

Yet for all its success in banishing hard times present, the war couldn't erase memories of hard times past. If anything, the war-driven boom appeared to confirm the correlation between war production and prosperity. Though our worker looked forward eagerly to victory over the Germans and Japanese, he also looked forward anxiously. His working life was nearly over, and between the new social-security system and the federal insurance that now safeguarded his bank account, he didn't worry greatly about himself and his wife. But for his children, he did worry. His daughter worked in an aircraft plant. What would she do when the government stopped ordering planes? His two sons had held only casual and temporary jobs before enlisting in the military in December 1941. Would there be work when the government mustered them out?

Not surprisingly, such problems occurred to members of the American government as well. Congress appointed a special committee on postwar economic policy and planning to find some solutions. Throughout the last year of the war, the committee heard testimony from hundreds of individuals drawn from all walks of public and private life. In November 1944, Dean Acheson, then assistant secretary of state for economic affairs, explained how matters appeared to the Roosevelt administration. Needless to say, Acheson didn't propose continuing the war as a means of maintaining prosperity. The correlation between war production and prosperity was strong, but not *that* strong. All the same, Acheson did recognize the requirement for active efforts to prevent the return of peace from precipitating a return to the misery of the years before the war. The assistant secretary pointed out to the economic-planning committee that the American economy must be made able not simply to sustain current employment but to absorb the 12 million persons now under arms. "If we do not do that," Acheson said, "it seems clear that we are in for a very bad time, so far as the economic and social position of the country is concerned. We cannot go through another ten years like the ten years at the end of the twenties and the beginning of the thirties"—a Democrat, Acheson saddled the Republicans with as much of the blame for the recent depression as possible—"without having the most far-reaching consequences upon our economic and social system."

Events abroad, namely the war, had pulled the American economy out of the last depression, and Acheson looked abroad for the way to keep the economy out of the next. The new markets afforded by government purchases had revived the economy during the war, but these almost certainly would shrink after the fighting stopped. New markets would have to be found overseas. Acheson reported to the committee, "No group which has studied this problem, and there have been many, as you know, has ever believed that our domestic markets could absorb our entire production under our present system. You must look to foreign markets."

Perhaps to drive his message home, Acheson remarked that under a "different system" the United States might manage to absorb its entire production domestically. Rising to the bait, one of the assistant secretary's interlocutors asked, "What do you mean by that?" Acheson responded, "I take it the Soviet Union could use its entire production internally." He added that accomplishing anything similar in America "would completely change our Constitution, our relations to property, human liberty, our very conceptions of law." No one wished to see this. "Therefore you must look to other markets, and those markets are abroad."

A questioner asked whether the United States wouldn't find itself competing in foreign markets against other countries that confronted similar difficulties of reconversion to peace. "I understand that problem," Acheson said. "But the first thing that I want to bring out is that we need these markets for the output of the United States." Once more: "We cannot have full employment and prosperity in the United States without the foreign markets."[7]

The Roosevelt administration had already taken steps to secure the foreign markets Acheson spoke of. At the Bretton Woods conference a few months earlier, American officials had persuaded most of America's allies to accept a new financial regime that fixed exchange rates among the world's leading currencies. Because of the strength of the American economy, this decision made the values of other currencies dependent on the dollar. The Bretton Woods conferees also agreed to support the creation of an International Bank for Reconstruction and Development (or World Bank) and an International Monetary Fund. The two organizations had the general purpose of channeling loans and investment to countries needing a hand up after the war. Here again America's economic clout allowed Americans to call the tune: providing most of the capital, Washington could specify who got how much and under what terms.

Acheson explained the essential terms to the committee. The American government, he said, desired to eliminate "all forms of discriminatory treatment in international commerce." The British, for example, must allow American goods to compete on an equal footing with goods from the various parts of the British empire. The American government was working for the elimination of quotas and other restrictions on imports, including tariffs, the better to allow penetration of foreign markets by American producers. Foreigners would gain reciprocal access to the American market, of course, but given the relative conditions of the economies of the major nations, Americans would get the better of the bargain for years to come. The American government was pushing foreign governments to curtail or dismantle cartels and to establish fair rules for dealing with state monopolies. The American government also sought expansion of opportunities for American investment overseas.[8]

Doing well and doing good

What Acheson described, and what Washington pursued, amounted to a remaking of the world political economy in the image of the American political economy. Like most missionary movements, this one combined substantial elements of self-interest and idealism. Reshaping relations abroad between individuals, government, and property on the model that obtained in America would benefit Americans economically by helping ensure that the end of the war not end the prosperity the war had brought to America. It would safeguard America's strategic position both by strengthening the liberal-capitalist countries and by circumscribing the sphere of socialism, and hence of socialism's Soviet sponsors. It would confirm American self-confidence as other countries ratified the choices Americans had made over several generations.

At the same time that an international reordering along American lines provided these benefits to the United States, it would equally serve the inhabitants of other countries—or so most Americans had little difficulty convincing themselves. Few Americans would have admitted to empire-building, certainly not in the customary sense. Empires, they thought, were what other great powers acquired. They generally interpreted America's imperialist fling of the turn of the century as a youthful indiscretion later righted by fair treatment of the offspring until majority, which conveniently was being reached—by the Philippines, anyway—at just this time. (That other child of the liaison, Puerto Rico, was closer to the American heart and therefore harder to release to the cruel world.) The 1898 oat-sowing aside, Americans abjured the sort of colonial imperialism that had marked the history of Britain, France, Germany, Italy, Japan, and various lesser powers. When America had expanded, as it undeniably had across North America, it brought acquired territory into the American union on terms equal to those enjoyed by the original thirteen states. If in the process it produced an empire, as some of its publicists during the 1840s and afterward granted, that empire was an empire of freedom and equality.

Though no one in the mid-twentieth century advocated adding European or Asian stars to Old Glory, the same principle of equality applied to

American thinking about overseas regions in the aftermath of World War II. Unlike the colonialist-imperialists of the nineteenth century, unlike the fascist-imperialists of the 1930s and early 1940s, and, as was just becoming evident, unlike the communist-imperialists of the late 1940s, Americans didn't look on foreign countries as territories to be conquered and held in subordination. Americans were no less expansive—imperialistic in their own way—than other great peoples, but their expansionism was, in theory at least, an egalitarian expansionism. (The practice would fall somewhat short, but practice always does.) With few exceptions, Americans believed that the spread of American institutions, notably democracy and capitalism, would bless the peoples who embraced them just as these institutions had blessed the American people. In offering democracy and capitalism to the world, Americans thought they were bestowing wonderful gifts.

They had a strong case. The merest glance at the rest of the globe in 1945 argued persuasively that the United States had more going for it than any other country. Its people were the most prosperous of any sizable group on the planet. They were the most secure in their homes and accustomed ways of living. They exercised the largest degree of personal liberty. By nearly every objective measure, a world that came to look more like the United States would be a world improved.

The improvement should start in Europe, the epicenter of the recent horrors, and it must begin with efforts to improve living standards. Americans noted that the depression of the 1930s had been even rougher on the Europeans than on themselves. It had ravaged the European economies and, worse, had created the political conditions that spawned fascism in Germany and Spain. Italian fascism, though dating from an earlier round of turmoil, tightened its grip during the 1930s. Throughout much of Europe, communist parties gained strength, though the fascists suppressed the left in the countries where the rightists held control. American leaders correctly interpreted the rise of both fascism and communism as protests against a status quo that had failed. After the war, they believed that preventing the protesters—now confined to the red end of the political spectrum—from regaining strength required preventing a return to the conditions that had promoted radicalism during the 1930s.

Unfortunately, prosperity for Europe seemed a distant hope at war's end. If wars are good for business in countries where they are *not* fought, they are terrible for business in countries where they *are*. After a decade of decline during the 1930s, when needed maintenance was often deferred, factories and transport networks across the continent were devastated during the war. Much of what the Germans hadn't destroyed in conquering France and the Low Countries in 1940 the Americans and British and French had ruined in liberating them in 1944. Britain's air aces had spared that country a ground war, but Hitler's bombers and rockets wreaked plenty of wreckage. Italy was still a shell, nearly two years after surrendering in 1943. Germany's cratered cities, blasted ports, and ravaged rail-yards displayed most clearly the ferocity of the fighting, although no one was wasting tears on the Germans at this point.

Nor did the situation improve during the months after the shooting stopped. Efforts to rebuild foundered on the lack of reconstruction capital. What scant reserves the depression hadn't depleted the war had quickly exhausted. American Lend-Lease aid had kept America's allies going against the fascists, but President Truman terminated the program abruptly on the fascists' capitulation. Hard weather in 1946 compounded the Europeans' hard luck. Bread rations shrank, tenants shivered, and without the esprit of war, morale plummeted.

The British moved first to break out of the dismalness. Realizing that only America had the money their country's reconstruction required, British officials trooped to Washington to get some. John Maynard Keynes and his fellows hoped to capitalize—literally—on good feelings left over from the war, but they discovered that these feelings stopped short of countenancing continued discrimination against American goods in the British empire. The British had been free traders in the days of Bright and Peel, when Britannia ruled the factories and countinghouses as well as the waves. Their enthusiasm for competition had waned as the competitiveness of second-generation industrializers like the United States had waxed. But the British needed the money, and in exchange for a concessionary credit of $3.75 billion, which the Americans insisted be spent on American goods, they agreed to begin dismantling the edifice of imperial preference.

The British loan proved a false start. It lacked "that little more of imagination, daring, and luck essential to success," as Dean Acheson later recalled. The loan didn't provide enough money to get the British economy going again, and it didn't even begin to address the problems faced by France, Italy, and the other European countries, which had less to pledge as collateral and generally poorer prospects of repayment. In addition, the loan made relations between the United States and Britain out to be a business deal. They were, but not a deal that met the normal standards of business practice. If they had, Keynes and company would have gone to New York rather than to Washington, to the House of Morgan rather than the House of Morgenthau. (In fact, Henry Morgenthau had retired before the arrival of the Keynes delegation, but the Treasury Department still bore the mark of the longest-serving secretary in its history.)[9]

The little more of imagination and daring Acheson referred to emerged during the spring of 1947. In April of that year, a coordinating committee of the American State, War, and Navy departments proposed a dramatic departure from previous policies. The committee projected American exports for 1947 to be about $16.2 billion. Against this amount, the United States would import $8.7 billion. The major force sustaining this overwhelmingly favorable balance of trade was American government spending, principally programs relating to the war. But these programs would soon begin to diminish sharply, leading the committee to declare, "The conclusion is inescapable that, under present programs and policies, the world will not be able to continue to buy United States exports at the 1946–47 rate beyond another 12–18 months." The shortfall of exports could have grave consequences. The committee noted that the president's economic advisory council was predicting a recession within the next year. "A substantial decline in the United States export surplus would have a depressing effect on business activity and employment in the United States. The net effect would depend on the direction and strength of other economic forces, but if the export decline happened to coincide with weakness in the domestic economy, the effect on production, prices and employment might be most serious."[10]

As Woodrow Wilson had done in 1915, when Britain and France ran out of cash to pay for American goods, Truman in 1947 moved to bolster Europe's purchasing power with American money. But this new bolstering included two significant differences. First, in 1915 private American banks had underwritten the purchases. In 1947 Truman proposed that the money should come from the United States Treasury. Second, in 1915 the American money had taken the form of loans. In 1947 Truman offered to give the money away, although not completely without strings. These two differences reflected two significant facts that separated the earlier period from the later. First, the Europe of the middle 1940s was a bad credit risk, as Keynes's coming to Washington demonstrated. Private American banks were hardly clamoring to plant their funds across the Atlantic. Second, and more important, the American government of the 1940s was far more willing than that of the 1910s to engage in large-scale intervention in economic affairs. The Republican party of Herbert Hoover had taken significant steps toward government-business collaboration during the 1920s. Franklin Roosevelt's New Dealers extended the corporatist coalition during the 1930s, adding organized labor and farmers, if somewhat de-emphasizing management. During the war, the dollar-a-year men almost erased the distinction between the public and private spheres from sizable portions of the economy.

As a consequence of all this, official Washington hardly blinked when Will Clayton, the undersecretary of state for economic affairs, returned from an inspection tour of Europe in May 1947 with a recommendation for a huge program of American aid. Clayton argued that the condition of the European economy was worse than the American government had known. And it was worsening further fast. Without prompt and massive American assistance, economic, political, and social disintegration would overwhelm the continent. "Aside from the awful implications which this would have for the future peace and security of the world," Clayton said, "the immediate effects on our domestic economy would be disastrous: markets for our surplus production gone, unemployment, depression, a heavily unbalanced budget on the background of a mountainous war debt. *These things must not happen.*"

To prevent their happening, Clayton advocated a program of some $20 billion of American assistance over three years. The aid should take the form of grants toward the purchase of American goods and services. In other words, the American government should pay American farmers, coal producers, shippers, and the like for goods and services to be delivered to Europe. This approach would lack a certain flexibility, since it wouldn't allow the Europeans to decide freely where the reconstruction money might best be spent. But it would most readily solve America's export problem.

Clayton specified another point that became central to American policy toward the continent. The United States, he contended, should insist that the European countries participating in the reconstruction program work toward economic—and perhaps eventually political—union. Belgium, the Netherlands, and Luxembourg had already moved in this direction, but the model that came most readily to the minds of American planners was that of the United States itself. If the original thirteen states hadn't buried their differences in the federal Constitution of 1789, North America almost certainly wouldn't have achieved the prosperity and relative peace it enjoyed in the subsequent century and a half. Although a United States of Europe remained almost beyond credible contemplation, the economic basis for such a result didn't. Clayton saw no alternative for the Europeans. "Europe cannot recover from this war and again become independent if her economy continues to be divided into many small watertight compartments as it is today."

Clayton encouraged the Truman administration to bring other governments into the act. The principal Western European nations—Britain, France, and Italy—must play major roles. Divided Germany posed special problems. Non-European countries with surplus goods could help. Yet Washington should keep this program out of the United Nations, which had its own programs, and where the Russians spoke with a decisive voice. "*The United States must run this show*," Clayton emphasized.[11]

Clayton's memo provided the basis for George Marshall's famous speech of one week later. With accustomed conciseness, the secretary of state outlined in a few hundred words the largest program of foreign aid in world history. As befitted the occasion—commencement at Harvard—Marshall stressed the idealistic aspects of the administration's proposal. "Our policy is directed not against any country or doctrine but against hunger, poverty, desperation, and chaos." Yet the anti-communist and anti-Soviet character of the plan peeked through. Marshall asserted that while the United States would help those governments prepared to assist in the reconstruction of Europe, "governments, political parties, or groups which seek to perpetuate human misery in order to profit therefrom politically or otherwise will encounter the opposition of the United States."

Marshall wisely set no price tag, and he postponed potential criticism by placing the immediate matter before the Europeans. "The initiative, I think, must come from Europe." Further, the Europeans must demonstrate an ability to pull together, as they so often had not in the past. "There must be some agreement among the countries of Europe as to the requirements of the situation and the part those countries themselves will take in order to give proper effect to whatever action might be undertaken by this Government." The secretary reiterated this requirement a few sentences later: "The program should be a joint one, agreed to by a number of, if not all, European nations."[12]

The Europeans—of both Western and Eastern camps—cooperated admirably. British foreign secretary Ernest Bevin caught Marshall's message on the fly and raced to make the plan a reality. Bevin called French foreign minister Georges Bidault and suggested a meeting of European countries to forge the sort of agreement Marshall required. Invitations soon went out and were accepted enthusiastically in the Western half of the continent. The Soviets declined, however, for both themselves and the countries in their Eastern European sphere. Moscow smelled a capitalist rat in the summons to European integration, and it refused to accept the kinds of market-freeing reforms the Americans were indicating they would demand. The Kremlin's veto averted a thorny problem: selling Congress an aid package that included assistance to the communists. It also kept costs down by confining the plan to half the continent.

Oedipus lex

Excluding the communists was all the more necessary in light of the other prong of the Truman administration's spring offensive of 1947. In March, the administration had decided to continue the process of supplanting Britain as the balancer of power in Europe and points east. Through most of the nineteenth century, the British had helped keep the peace of Europe by preventing any one country or group of countries from getting too powerful. They fought with Prussia and Russia against France at the beginning of the century. By mid-century, they were fighting with France against Russia. The century's end found them allied once more with the Russians and still with France against Prussia and the rest of the Germans. The Anglo-French-Russian coalition didn't suffice to balance German power during World War I, and the United States was obliged to step in to prevent the framework of European security from toppling over, as well as to secure American creditors against default by Britain and France. Americans hoped the rescue was a one-time operation, and they spent the interwar period attempting to deny evidence that the unification of Germany had permanently upset the continental balance. The evidence grew overwhelming by 1940, and shortly thereafter America's balancing forces again traversed the Atlantic.

World War II unweighted Germany considerably, but it also lightened the countries to Germany's west. The new heavyweight of the continent was Russia. American leaders, having twice discovered the costs of tardiness in righting a tipping continental balance, didn't intend to procrastinate again. At least until the European democracies regained their feet, the United States offered the only significant offset to the Soviets.

Unfortunately for Washington's prospective Palmerstons, balance-of-power geopolitics carried no cachet with the American public. Rather, it smacked of the intriguing and double-crossing that Americans had learned to associate with the chancelleries of the Old World during America's days of innocent insulation from European affairs. What was obvious in the corridors of the State Department and the Pentagon was far from it in the factories of Akron and the grocery stores of Sacramento.

If the Truman administration were to convince the country that securing American interests required continuing attention to the geopolitical balance in Europe, the administration needed a galvanizing issue on which to make its case. In the 1850s, a quarrel over the holy places in Palestine had served as the catalyst for the Crimean War, in which Britain joined with France and Turkey to counter the Russians. Nearly a century later, the catalyst surfaced slightly west of Palestine but still in that turbulent region where the Balkans meet the Middle East.

In February 1947, the British decided to divest themselves of responsibility for maintaining a pro-Western status quo in Greece and Turkey. Yet before dropping this self-assumed burden, they offered Washington first refusal. The Soviets had been pressuring Turkey since the end of the war to gain greater influence in the Turkish Straits, and although the Turks to date showed no sign of buckling, American officials feared that the psychological and economic strain of protracted mobilization might soon begin to soften Turkish resistance. General Dwight Eisenhower, writing for the American joint chiefs of staff, summarized the situation: "The danger remains that Turkey, unless given positive assurances including concrete assistance, might so interpret the possibilities of the future as to yield to Soviet pressure short of direct military measures." Eisenhower went on to assert that the rest of the Middle East took Turkey as a test of Western willingness to block Soviet expansion. If the Russians succeeded in dominating Turkey, he said, it would be "highly probable that all the Middle Eastern countries would then come rapidly under similar Soviet domination." Such an occurrence would doubly endanger the West: once by loosing the Russians in the warm waters of Indian Ocean, twice by allowing them a hand on the spigot to Europe's oil supply.[13]

But aid to Turkey lacked the popular appeal needed to set American policy on a markedly different course. Few Americans knew or cared much about Turkey. Most of those few associated the Turks with the wrong side in Richard the Lion-Hearted's crusade of the second century of the present millennium, or Woodrow the Righteous's crusade of the second decade of the present century. In between, some Americans recalled, the Turks had beaten up the Greeks and other hapless peoples unfortunate enough to fall under the sultan's sway.

Greece—now *there* was a thought. Americans had shown a soft spot for the birthplace of democracy since Thomas Paine and Patrick Henry had begun making democratic noises in the eighteenth century. Sometimes the sturdiness of Sparta appealed more to Americans than the erudition of Athens, but whatever one's pick in the Peloponnesian War, Americans trained in the classics—which was to say nearly all those who participated in the political debates of the revolutionary and early national period—felt themselves not far removed from Hellas's golden age. During the 1820s, Byronic types in the United States agitated for American aid to assist the Greek rebellion against the unspeakable Turks. Though John Quincy Adams ignored the appeals, favoring instead the noninterference provision of his and James Monroe's eventually famous doctrine, the experience reinforced for many Americans the feeling of connection to Greece.

In the late 1940s, Greece again faced danger. This time the danger was largely internal, in the form of a civil war of uncompromising ferocity. (Some things hadn't changed much since Thucydides' day.) Yet amid the exigencies of the developing conflict between East and West, the distinction between internal and external was eroding rapidly. Without a doubt, the Greek rebels, communists mostly, were receiving help from co-ideologists in Yugoslavia, Albania, and Bulgaria. Since the communists of those countries spoke kindly of the Kremlin, the inference reasonably followed that a communist victory in Greece would strengthen Moscow's position in the Mediterannean.

If the world had been a neater place than it was in the aftermath of its biggest war ever, the beleaguered government in Athens would have possessed the political virtues Americans held dear. It might have been as likable as the constitutionalist cause in Spain during the civil war there a decade before. But the Greek government wasn't the sort to inspire Abraham Lincoln brigades. An American official, sent to Athens to assess the government's prospects for gaining control of the country's crumbling economy, returned shaking his head. "There is really no state here in the Western concept. Rather we have a loose hierarchy of individualistic politicians, some worse than others, who are so preoccupied with their own struggle for power that they have no time, even assuming capacity, to develop economic policy." This official said he lacked hard proof of venality in high places, but circumstantial evidence was discouragingly persuasive. The civil service was a "depressing farce." Government activities were distinguished by an "almost complete deterioration of competence."[14]

Yet if the previous ten years had taught American leaders anything, it was that in a pinch one takes allies as one finds them. Americans had gladly accepted the help of Stalin, hardly an angel, against the greater devil Hitler. Now that Hitler's demise had led to Stalin's promotion to diabolical pre-eminence, the deficiencies of the Greek government rated hardly a quibble.

Consequently, when Britain proposed to pull the plug on Athens, American officials moved quickly to keep the current flowing. The State, War, and Navy departments huddled to prepare a unanimous recommendation for the president. The proposal cast the current contest in Greece in apocalyptic terms. The Greek civil war, it said, was nothing less than a continuation of the planet-determining struggle recently waged against the fascists. "A cardinal objective of United States foreign policy is a world in which nations shall be able to work out their own way of life free of coercion by other nations. To this end the United States has just finished fighting a war against Germany and Japan, who were attempting to impose their will upon other nations. To the same end, the United States has taken a leading part in establishing the United Nations, which is designed to make possible freedom and independence for all of its member nations." American policy had as its aim resistance to aggression and to the imposition by outside forces of dictatorial regimes, "whether fascist, nazi, communist, or of any other form."

The joint report went on to separate the sheep from the goats. "There is, at the present point in world history, a conflict between two ways of life. One way of life is based upon the will of the majority, free institutions, representative government, free elections, guarantees of individual liberty, freedom of speech and religion, and freedom from political oppression. The second way of life is based upon the imposition of the will of a minority upon the majority, upon control of the press and other means of information by the minority, upon terror and oppression." The committee dismissed as immaterial the differences in objectives between communists and fascists, since the crux of the world struggle had little to do with objectives. "The major issue that is posed for the world is not one of objectives, not one between socialism and free enterprise, not one of progress or reaction, not one of left versus right. The issue is one of methods: between dictatorship and freedom; between servitude of the majority to a minority and freedom to seek progress."

The report concluded with an appeal to maintain the current geopolitical balance. Greece afforded a test of American willingness to take action against "a unilateral gnawing away at the status quo." If unresisted in Greece, the gnawing would eat into American security. "The national security of the United States depends to a large degree on the maintenance of the principles of the United Nations and on maintaining the confidence of other nations in these principles. A seizure of power by a Communist minority in Greece would seriously impair that confidence."[15]

This report was noteworthy in three respects. First, by conflating communism with fascism, it simplified enormously the task of mobilizing public opinion against the Soviet Union. If Stalin was simply Hitler with a better mustache, Americans had merely to transfer to the former their demonstrated loathing for the latter. Second, by dichotomizing the world into the realms of "dictatorship and freedom," the report rendered unnecessary a defense of the Greek government on that government's mostly missing merits. The regime in Athens might be brutal, corrupt, and inefficient, but it wasn't a dictatorship, and so by process of elimination it landed in the ranks of the free. Third, by asserting the necessity of instilling confidence among the non-communist nations in America's determination to safeguard the status quo, the report placed the importance of Greece beyond disproof. Was Greece strategically significant? Did the government there deserve saving? Were the Greek rebels really on the Kremlin's payroll? These questions faded into irrelevance once Greece became a proving ground for American resolve.

The report succeeded smashingly. Within days, Truman embedded its arguments and conclusions in a speech given before a joint session of Congress. The president echoed the report in declaring that the struggle in Greece was an extension of the late conflict against Germany and Japan. He lifted almost unchanged the words of the report proclaiming an essential antagonism between two ways of life, one based on the "will of the majority" and distinguished by "free institutions, representative government, free elections," and so on, and the other based on the "will of a minority forcibly imposed upon the majority" and relying on "terror and oppression, a controlled press and radio," and so forth. He emphasized the crucial role of confidence as he described the "profound effect" a communist takeover in Greece would have on other countries. "Discouragement and possibly failure would quickly be the lot of neighboring peoples striving to maintain their freedom and independence."

Truman declared that in order to ensure victory in the abiding conflict against dictatorship, to assist those striving for freedom, and to reinforce the confidence of countries looking to American leadership, it "must be the policy of the United States to support free peoples who are resisting attempted subjugation by armed minorities or by outside pressures." To these ends, the president requested $400 million in aid for Greece and Turkey.[16]

Truman's message entered an arena of public debate already primed for the ideas he delineated. The equating of communism with fascism drew on analogies made by conservatives during the late 1930s and especially during the period of German-Soviet collaboration between 1939 and 1941. The second half of the war weakened the conceptual axis linking Berlin to Moscow, but as the Grand Alliance came apart after 1945 conservatives reverted to form, and others joined them. Just weeks before Truman's address, journalist and pundit Herbert Matthews wrote that "fascism is not dead." It had simply looped around from the right wing of politics to the left. "You pay your money and you take your choice—a dictatorship of the Left or Right, Communism or Fascism." Matthews asserted that American confusion and dismay at the tension that had marked international affairs since the war resulted from a basic wartime misconception. "We had been trying desperately to fool ourselves into the belief that Fascism and Communism were enemies because of fundamental, systematic differences. Instead, they represented two factions within the same camp." Now that Hitler was dead, the Stalinist faction rated top billing. But the difference was hardly worth bothering about. "It is really a matter of labels."[17]

Not everyone agreed with Matthews. Leftover liberals like Henry Wallace saw a sizable distinction between socialism, which, while imperfect in practice, in principle exalted equality and social uplift and appealed to humanity's better side, and fascism, which built unashamedly on the exclusionary, destructive, racist rantings of the Nazis and their fellow-travelers. Wallace suggested that Moscow's activities in Eastern Europe, though hardly commendable, represented not an aggressive intent to conquer the continent, but a defensive desire to avert a repetition of the disasters that had befallen Russia twice in thirty years.

Such subtleties, however, were largely lost on Americans searching for an answer to the vexing question of why peace hadn't followed victory. During the war, Hitler had provided an emotion-satisfying and emotion-intensifying focus of evil, capable of stirring the American people to the sacrifices the war required. The fact that Hitler was an undeniably despicable character made him all the more convincing as the cause of the world's misfortunes. Predictably, Americans came to believe that once the Allies destroyed Hitler—and his Japanese accomplices—most of the planet's pressing problems would cease.

But troubles persisted beyond Hitler's defeat. Why? Two possible answers presented themselves. First, Hitler had not been the chief source of the recent disaster. Second, though Hitler had died in that Berlin bunker, his specter lived on in the Soviet Union. Accepting the first answer required rethinking most of a decade of world history. It also required figuring out what *had* caused the horrors just past and the difficulties still continuing. The second answer left prior explanations intact, and it allowed an easy transfer of hostile emotions from Hitler to Stalin—a transfer, moreover, that facilitated the policies American leaders judged necessary to stabilize America's international position. The conversion of Stalin from ally to enemy could be expected to produce a certain amount of cognitive dissonance. Yet as poets, novelists, and playwrights have long known, hatred follows more readily from trust betrayed than from indifference, and in this regard Stalin's wartime cooperation actually facilitated his re-identification as Hitler's heir.

Treating communism as "red fascism" served an additional purpose. It provided a persuasive rationale for thoroughly overturning the longest-established and most cherished principle of American international relations: nonentanglement in the peacetime affairs of foreign countries. By itself, the Truman Doctrine didn't represent a repudiation of the teachings of the Founding Fathers. Neither Washington nor Jefferson had expected the United States to hold aloof from the world. The America of their day was a country actively engaged in the pursuit of profit and other American interests, from Shanghai to Tripoli. And certainly neither the commanding general in the War for American Independence nor the commander-in-chief who sent American warships against the Barbary pirates thought the United States should retire meekly from international competition. What they objected to—Washington in declaring that "it is our true policy to steer clear of permanent alliances," and Jefferson in calling for "peace, commerce, and honest friendship with all nations, entangling alliances with none"—was a continuing attachment of the United States to one or more foreign countries. "Permanent" or "entangling" alliances would diminish America's freedom of action by binding the American government to other governments and by obliging the American people to pursue interests of other peoples' choosing. Such a policy the United States must forswear.[18]

Yet this was precisely what the Truman administration had in store for the country. Legally speaking, the North Atlantic alliance didn't follow from the Truman Doctrine. On the contrary, it was an entirely different breed of constitutional cat. The Truman Doctrine, in the tradition of the Monroe Doctrine, was nothing more than a unilateral statement of American intent, issued on the authority of the president and revocable at the president's discretion. In practice, of course, Truman required congressional concurrence in funding his policies, and he would feel political constraints in the event he decided to backtrack in the future. Nonetheless, there existed a qualitative difference between an executive decree like Truman's statement and a treaty of alliance, which required the approval of the Senate and legally bound the United States as a nation.

But though not legal kin to the Truman Doctrine, the Atlantic alliance followed both geopolitically and ideologically from Truman's declaration. Geopolitically, if Greece and Turkey deserved defending, then Western Europe deserved defending more. A Soviet breakthrough into the Mediterranean and the Middle East threatened the United States because such a breakthrough threatened Europe. Ideologically speaking, the United States was rushing to the aid of Greece and Turkey on the reasoning that communist aggression, being operationally equivalent to the fascist aggression of the previous decade, responded only to the kind of unequivocal demonstration of steadfastness the democracies had failed to provide in the earlier case. In this regard, an American alliance with the countries of Western Europe was simply another way of saying what Truman had already said in his doctrine.

Regardless of its connections to the Truman Doctrine, the North Atlantic treaty of 1949 struck many in the United States as a significant departure from American traditions. Foremost among the oppositionists stood Robert Taft, the Republican senator from Ohio. Taft didn't object to a large role for the United States in the world. Nor did he complain about an American commitment to defend Europe against the Soviet Union. What he decried was the manner in which the Atlantic treaty tied the American people to the actions of governments over which they had almost no control. "By executing a treaty of this kind," Taft told his colleagues in the Senate, "we put ourselves at the mercy of the foreign policies of eleven other nations, and do so for a period of twenty years." At present, the foreign policies of the countries in question didn't conflict with those of the United States. But who knew how long this parallelism would last? "The government of one of these nations may be taken over by the Communist Party of that nation." Because the treaty bound nations and not particular parties, this would place the United States in an intolerable fix.

As an alternative to a treaty of alliance, Taft favored a presidential declaration on the order of the Truman Doctrine—although, unwilling to give credit to the Democratic president, he used the example of the Monroe Doctrine. "I am in favor of the extension of the Monroe Doctrine to Western Europe," he said. The crucial advantage of such a course was the freedom it allowed the United States to act as the American people, not the people or governments of other countries, saw fit. "The Monroe Doctrine was a unilateral declaration. We were free to modify it or withdraw from it at any moment." Under the Monroe Doctrine, the American people determined what to do in case of war. "We were free to fight the war in such a manner as we might determine, or not at all." On each of these counts, the Monroe Doctrine excelled the Atlantic treaty.

Beyond its other deficiencies, Taft said, the Atlantic pact presaged American efforts to arm the Europeans. The Truman administration denied any commitment to send weapons to Europe. But Taft contended that such a commitment was the nearly necessary corollary to the treaty. "I think the pact carries with it an obligation to assist in arming, at our expense, the nations of Western Europe." Not only would this obligation unconscionably burden the American taxpayer, it would promote an arms race that would make the world more hazardous. The treaty's signers might call their creation a defensive alliance, and so it was. But the Russians might not interpret it so. They would certainly build up their own forces, and they might decide to attack rather than wait for the West to get stronger.

Where would it all lead? "The history of these obligations has been that once begun, they cannot easily be brought to an end," Taft said. After Western Europe, what next? "If the Russian threat justifies arms for all of Western Europe, surely it justifies similar arms for Nationalist China, for Indochina, for India, and ultimately for Japan; and in the Near East for Iran, for Syria, and for Iraq. There is no limit to the burden of such a program or its dangerous implications."[19]

Taft received a standing ovation in the Senate chamber at the conclusion of his speech, but on the roll call only thirteen senators joined him in refusing assent to the treaty. With the Senate's acceptance of the Atlantic pact, the generation of World War II fundamentally altered the course of American history, killing at last the dream of the Revolutionary War generation that the United States might remain free of the quarrels that seemed perpetually to embroil the Old World.

Ironically, the creation of the Atlantic alliance also dealt a grievous blow to the dream of American internationalists, whose views were the polar opposites of the Founding Fathers'. Since before World War I, the internationalists had argued for a scheme of world government, or at least world law, that would bring order to the anarchy of international affairs, and thereby render war obsolete as a means of solving disputes between nations. Their hopes soared with the 1919 approval of the League of Nations by the Paris peace conferees, but then plunged when the Senate rejected the Versailles treaty. During the interwar years, the internationalists' message fell on stony ground, only to sprout once more in the blood-fertilized soil of the second great war.

Disillusionment came slower this time. Though the Yalta conference essentially guaranteed that the postwar settlement would be based on a spheres-of-influence arrangement rather than a genuine collective-security approach, Yalta's implications took many months to make themselves evident. Meanwhile, the United Nations began work amid great enthusiasm for a new day in human affairs. Modest success in such jobs as ending Britain's Palestine mandate and finding acceptable terms for a ceasefire between India and Pakistan in Kashmir afforded encouragement.

But the hardening of positions on either side of the Cold War increasingly indicated a marginal future for the United Nations and the internationalist spirit it represented. The Truman administration's insistence on making the Marshall Plan an American show demonstrated Washington's unwillingness to look to the United Nations for guidance in reconstructing Europe. The Truman Doctrine signaled a refusal to rely on the United Nations to solve the problems of Greece and Turkey. The decision for the Atlantic treaty, by demonstrating yet again America's lack of confidence in internationalist schemes and by making explicit Washington's preference for geopolitical guarantees of global security, pushed the internationalists still further into the background.

Robert Taft got it right on this point, as on several others in his critique of the Atlantic treaty, when he declared that the approach embodied in the alliance violated "the whole spirit of the United Nations Charter." Taft added, "The Atlantic Pact moves in exactly the opposite direction from the purposes of the Charter and makes a farce of further efforts to secure peace through law and justice. It necessarily divides the world into two armed camps. It may be said that the world is already so divided, but it cannot be said that by enforcing that division we are carrying out the spirit of the United Nations."[20]

Red tides and monoliths

If events in Europe weakened American internationalism, developments in Asia finished it off. During the war, many American officials, led by Franklin Roosevelt, had hoped that ally China would step into the power breach soon to be created by the defeat of enemy Japan. This hope never quite matured into expectation, on account of the civil war between the Communist party of Mao Zedong and the Nationalist government of Chiang Kai-shek that formed the background, and often the foreground, to the anti-Japanese war. After the world war ended, Truman sent George Marshall to China to try to negotiate a settlement of the Chinese conflict. When this failed, Washington backed Chiang in an effort to secure a settlement by the force of Nationalist and American arms. This also failed, however, and through the beginning of 1949 the inept and unpopular Nationalists lost ground with accelerating speed to the Communists. The vocal pro-Nationalist lobby in the United States called for sending American troops to assist Chiang, but Truman, convinced that the Nationalists had passed beyond help, refused. During the first half of 1949, he tried to cut his losses, gradually disengaging from Chiang. In August of that year, the State Department published a sua culpa, blaming Chiang's defeat on Chiang, where in fact the bulk of the blame belonged.

Yet the Truman administration now found itself hoist with its own petard. Having divided the world conceptually into two camps, Truman left no third category for independent communist movements like Mao's. By no stretch of any but the most fevered imagination was Mao a puppet of the Kremlin. Stalin distrusted Mao as he distrusted all communists who didn't owe their positions to him. Even so, the (ideo)logic of the administration's policies required that Mao's victory be judged a victory for the Soviet Union, and a defeat for the United States.

Truman's Republican rivals, still stunned at their come-from-ahead defeat in the presidential and congressional races of 1948, gladly judged it so. The overconfident Thomas Dewey had deliberately downplayed foreign affairs in his campaign, figuring he would need Democratic support when he became president. The now-smarter but yet-smarting Republicans determined not to make the same mistake again. The communist conquest of China afforded the GOP a perfect club for beating Truman and the Democrats, and during the rest of Truman's tenure they pounded the president unmercifully.

The pounding had some curious but not unpredictable effects on the administration's policy for East Asia. A decisive factor in Truman's refusal to send troops to help Chiang was the well-founded belief on the part of the president and his top military advisers that a war on the Asian main-

land would swallow American forces to no good purpose. Postwar demobilization had shrunk the United States army from more than eight million
soldiers in June 1945 to less than one million two years later. Occupying
Germany and Japan occupied many of those still in uniform. As a consequence, the troops simply weren't available for a major effort in China.
Even if they had been, their numbers would not nearly have matched
those of the People's Liberation Army, which additionally would have
possessed a home-field advantage.

Rather than draw a line in China or elsewhere on the East Asian continent, American leaders initially chose to establish a defensive perimeter
along the chain of islands offshore, from the Aleutians in the north
through Japan and the Ryukyus, including Okinawa, to the Philippines
and Australia and New Zealand in the south. The offshore strategy played
to American strength in two respects. First, it concentrated on territories
where the United States already had a foothold: by ownership (the Aleutians), occupation (Japan and the Ryukyus), long-term base agreements
(the Philippines), and the existence of like-minded governments (Australia
and New Zealand). Second, it took advantage of America's naval and air
superiority. Even after demobilization, the United States still boasted the
world's largest navy, and alone among the world powers its bombers
packed atomic thunderbolts.

Critics of the administration's China policy denounced the offshore
strategy as simply a cover for abandoning Chiang and the Nationalists. All
the same, as no less an advocate of a staunch policy against Asian communism than Douglas MacArthur noted, the offshore strategy represented
not a retreat but a signal advancement of American power. "Our defensive
dispositions against Asiatic aggression used to be based on the west coast
of the American continent," MacArthur declared from his occupation
headquarters in Tokyo. "The Pacific was looked upon as the avenue of
possible enemy approach. Now the Pacific has become an Anglo-Saxon
lake and our line of defense runs through the chain of islands fringing the
coast of Asia."[21]

But military considerations told only half the story, and the less informa-
tive half at that. The communist victory in China sensitized the Truman
administration to charges—which it had invited upon itself by its two-
worlds, zero-sum philosophy of international affairs—of capitulationism
in Asia. Whether or not the forces of democracy in Asia (whatever those
might be) could stand further communist encroachments, the forces of
Democracy in Washington (namely the president and his party) couldn't.
Further, by emphasizing American credibility as the crucial factor in democracy's survival, the administration had converted the regional strategic
problem of Asian defense into a global psychological problem. Generals
regularly pull back from exposed salients to consolidate their units and
entrench or prepare to counterattack. Truman couldn't do likewise without raising doubts among the Atlantic allies and numerous other observing

nations about America's willingness to stay the course Washington had set for itself.

As a result, at the next opportunity to get involved in an Asian civil war—this time in Korea—the Truman administration leaped into the fray. As soon as Kim Il Sung's communist North Korean army struck against the South Koreans of Syngman Rhee in June 1950, Truman ordered American forces to the rescue. Truman's move made little military sense. If the Korean fighting was a prelude to a broader communist offensive, as many in Washington suspected, then the United States should reinforce its position in Europe and Japan rather than stretch its capabilities thinner. If, on the other hand, the Korean affair was basically a tussle among Koreans, then there was no good reason to sacrifice lots of American lives to ensure the victory of one faction over the other.

But Truman's decision was based on political and psychological factors, not on military ones. Politically, the Republican drumbeat continued, louder than ever after January 1950 when Joseph McCarthy had taken charge of the band, waving his elusive lists of equally elusive State Department subversives. Congressional elections loomed in the fall, and Truman's Democratic colleagues didn't relish running on the record of an administration that had failed a second time in quick succession to prevent a communist victory in Asia.

Psychologically, Korea became the new Munich. The democracies had waffled before Hitler in 1938, and world war had followed. If they waffled again, the result would be similar, only worse in the age of atomic weapons. Korea's strategic significance was beside the point. Neither had the Sudetenland been strategically vital. The important issue was whether the United States and the other countries of the Free World would stand up for the principles they espoused.

Resting as it did on the dualistic world view of the Truman Doctrine, the American response to the fighting in Korea also reinforced that world view. As happens in almost every war, once American soldiers started dying, support for the war developed a dynamic all its own. Fathers and mothers, wives and children of the fallen needed to believe that their loved ones had perished in a worthwhile cause. The autocratic Syngman Rhee hardly qualified as such a cause. Neither did South Korea, about which most Americans knew next to nothing. Why should the United States defend one gang of Koreans against another? What did Rhee have to offer that was so much better than what Kim promised?

On the other hand, if the conflict in Korea was really the work of the Kremlin, then drawing the line in Korea made sense. If the Russians had been behind the trouble in Greece and Turkey in 1947, they certainly could be behind the trouble in Korea now. It would be just like those Reds to test America's resolve in an area that had little intrinsic importance to the United States. That was exactly how Hitler had started.

At a more objective level as well, the Korean fighting reinforced American perceptions of a monolithic communist conspiracy against world peace. Less from solidarity with the Soviets than from fear of the United States, China in February 1950 signed a treaty of friendship with Moscow. What form friendship would take remained unclear during the next several months, but the mere fact of the treaty put Americans on alert for signs of collusion among the communists. At about the same time as the signing of the Moscow-Beijing accord, Stalin's delegate to the United Nations began a boycott of the international body to protest its refusal to seat Beijing's representative. Under these circumstances, the North Korean attack a few months later appeared more than coincidental.

The Moscow-Beijing-Pyongyang linkage strengthened during the next half-year. In the first phase of the fighting, the Chinese and Soviets openly provided diplomatic and logistical support to their Korean comrades. La- ter and less openly, Chinese troops fought alongside the North Koreans as "volunteers." The Chinese charged the United States with using the Ko- rean conflict as a pretext for reintervention in China's civil war. The charge had merit: Washington's refusal to recognize Beijing demonstrated a desire to restore Chiang to power, or at least delay his final defeat. So did Truman's decision, simultaneous with the dispatch of American soldiers to Korea, to deploy American naval vessels between the Chinese mainland and the Nationalists' last redoubt on Taiwan. When American troops approached the Chinese-Korean border in the autumn of 1950, amid calls from American conservatives for smashing the Chinese revolution, Beijing not unnaturally feared for its safety. The Chinese entered the war in huge numbers in November, thereby closing the circle of cause and effect in the matter of communist solidarity and American opposition. Moscow ap- plauded the Chinese move, and when Truman rumbled about using atomic weapons against the Chinese, the now-atomically-armed Russians rumbled back.

Beyond reinforcing American perceptions of a unified communist movement—partly by in fact reinforcing communist unity—the Korean War also drove the last nail into the coffin of the idea that the United Nations might serve as a significant force for peace. The Soviet boycott enabled the Truman administration to gain Security Council acceptance of a resolution calling on the members of the United Nations to lend support to South Korea against the communist invaders. By this stroke, the admin- istration managed to wrap its activities in Korea, which it surely would have undertaken anyway, in the blue and white flag of the international organization.

At the time, the stroke seemed a stroke of good fortune for both the United States and the United Nations. The United States gained the support of fourteen other member countries that sent contingents to Ko- rea, and, more important, it gained the approval of the United Nations for its basic approach to world affairs. The United Nations achieved unprece- dented stature as a vital participant in an issue of undeniable importance.

But this very stature destroyed what remained of the United Nations' ability to act as a disinterested arbiter of international disputes. By becoming a belligerent in a conflict separating the superpowers, the United Nations utterly lost its credibility with that one-third of humanity now governed by communist regimes. Even neutralists like India looked askance at the manner in which the Americans had succeeded in politicizing the United Nations. To much of the world, the organization seemed simply another American agent in the Cold War.

In time, Americans would decry the politicization of the United Nations. They would express particular outrage at the numerous anti-American resolutions passed by Third World majorities in the General Assembly. There was no little irony in such complaints, since the United States, following the Soviet Union's return to the Security Council, originated the use of the General Assembly as an instrument for circumventing vetoes in the smaller body. In September 1950, Dean Acheson presented to the General Assembly a "Uniting for Peace" resolution allowing precisely such end runs, and in November the assembly, then dominated by a pro-American majority, accepted the measure. Looking back from the late 1960s, Acheson remembered pondering the chances that the scheme might backfire. "But present difficulties outweighed possible future ones," he remarked, "and we pressed on."[22]

By using the United Nations as a hammer against the communists, the Truman administration simultaneously finished off the dream of American internationalism (for forty years, anyway). The United Nations, of course, continued to exist, and it occasionally sent peacekeeping forces to supervise the separation of warring factions in regional disputes. Its social-service agencies did good work despite an inevitable tendency toward bureausclerosis. The General Assembly operated as a forum where small countries could feel important. But in the really serious matters of global security, for four decades after the beginning of the Korean War, the United Nations played no significant part.

The marginalization of the United Nations transformed the internationalist ideal of collective security into a truncated system of selective security. In this new system, the superpowers looked not to the world organization for protection, but to themselves and their allies and clients. The framework of global security that emerged during the 1950s bore a striking resemblance to the alliance arrangements of the period before World War I—which were precisely the arrangements the pioneer internationalists who had sponsored the League of Nations had worked so hard to relegate to history.

★

CHAPTER TWO

The National Insecurity State 1950–1955

Policy as blunt instrument

The death of American internationalism left only the balance-of-power approach to global security. An influential school of political theorists, led by Hans Morgenthau, contended that no genuine alternative had ever existed. Internationalism, said Morgenthau, was a dangerous chimera. Nations sought power, ceaselessly and without limit. Only the balancing forces of other nations, likewise seeking power, prevented the unruliness endemic to international relations from exploding into bloody anarchy. Conscious choice hardly entered the picture. It was the balance of power, or nothing.

Whether or not Morgenthau and his fellow "realists," including George Kennan, the principal propagandist for the Truman administration's policy of "containing" the Soviet Union, were right, by 1950 American international relations rested squarely on a geopolitical, balance-of-power philosophy. They rested, as well, on the ideological, two-worlds premise of the Truman Doctrine, which held that a victory for communism anywhere was a defeat for noncommunism everywhere. The two approaches—the geopolitical and the ideological—meshed conveniently in the near aftermath of World War II. The United States and the Soviet Union were the greatest of the great powers, and they were also the primary exponents, respectively, of democratic capitalism and communism. What benefited the United States geopolitically benefited, by and large, the ideology America embraced; and vice versa. What benefited the Soviet Union as a country benefited the cause the Soviet Union represented, and vice versa.

Between its balance-of-power geopolitics and its Manichean ideology, the Truman administration laid out an enormous task for itself. The United States must balance the Soviet Union militarily and strategically, and at the same time it must oppose communism wherever the specters of Marx and Lenin reared their ugly heads. The work had begun in Europe, but now was most demanding in Asia. American troops were fighting Chinese and North Koreans to keep communism, and presumably Soviet power, from spreading down the Korean peninsula. American ships were cruising the Chinese coast to prevent communism from leaping to Taiwan. American weapons and military advisers were arriving in Indochina to stop the communist Viet Minh from taking over France's Southeast Asian colony.

Where American troops, ships, or weapons might have to go next, none could say. But the Truman administration prepared for the worst. In fact, the preparing predated the outbreak of fighting in Korea. During the spring of 1950, Truman's National Security Council had developed a blueprint for American policy in the Cold War. The blueprint, denominated NSC 68, began with an outline of the shifts in the balance of power that had led to the current confrontation between the United States and the Soviet Union. The period since 1914 had witnessed the collapse of five imperial systems—Ottoman, Austro-Hungarian, German, Italian, and Japanese—and the enervation of two more—British and French. During the nineteenth century, the various empires had held each other in check. No longer. "During the span of one generation, the international distribution of power has been fundamentally altered," the NSC paper declared. In place of the several centers of power, there now existed only the United States and the Soviet Union. "Power has increasingly gravitated to these two centers." Since the Red Army had begun pushing west after lifting the German siege of Stalingrad, the Soviets had gathered vast territories and populations into their sphere. This had required balancing action by the United States, and still did. "Any substantial further extension of the area under the domination of the Kremlin would raise the possibility that no coalition adequate to confront the Kremlin with greater strength could be assembled."

Yet traditional balance-of-power considerations explained only half the current struggle. The ideological gulf between the United States and the Soviet Union gave the geopolitical rivalry unprecedented urgency. Paraphrasing Truman's March 1947 speech, the authors of NSC 68 described "a basic conflict between the idea of freedom under under a government of laws, and the idea of slavery under the grim oligarchy of the Kremlin." The authors carried the dichotomizing much further than Truman had. On one hand, they spoke glowingly of the "marvelous diversity," the "deep tolerance," and the "lawfulness" of a free society. Such a society attempted to foster and maintain "an environment in which every individual has the opportunity to realize his creative powers." In contrast to this noble vision, the authors depicted an alternative society that had fallen under "the domination of an individual or group of individuals with a will to absolute power." Nothing could be more forbidding. "Where the despot holds absolute power—the absolute power of the absolutely powerful will—all other wills must be subjugated in an act of willing submission, a degradation willed by the individual upon himself under the compulsion of a perverted faith. It is the first article of this faith that he finds and can only find the meaning of his existence in serving the ends of the system. The system becomes God, and submission to the will of God becomes submission to the will of the system."[1]

There was more along these lines, pages and pages of it. Even assuming the authors thought they were providing accurate accounts of life in the United States and in the Soviet Union (Was American society genuinely fostering an environment in which African-Americans, for instance, could realize their creative powers? How many Soviet citizens honestly found the meaning of their existence only in serving the Stalinist system?), what was the purpose of it all? Who were the authors preaching to? NSC 68 was a top-secret document that would remain classified for almost a quarter-century. Only persons near the center of the policy process would read it. Presumably, such persons didn't need to be convinced of the superiority of the American way of life over that of the Soviets.

Dean Acheson later explained the exercise. "The purpose of NSC 68 was to so bludgeon the mind of 'top government' that not only could the President make a decision but that the decision could be carried out." Acheson added, "The task of a public officer seeking to explain and gain support for a major policy is not that of the writer of a doctoral thesis. Qualification must give way to simplicity of statement, nicety and nuance to bluntness, almost brutality, in carrying home a point."[2]

The aim of this bureaucratic brutality, as it turned out, was to achieve a huge increase in the American defense budget. For this, the president and other members of top government had to be not merely reminded of the Soviets' evil intentions, but scared witless. "The Soviet Union is seeking to create overwhelming military force," the NSC 68 authors asserted. Moscow had added atomic armaments to its huge conventional forces, and American intelligence experts predicted that by 1954 the Soviets would possess as many as 200 atomic bombs. For the first time in American history, an enemy would possess the capacity to strike quickly and devastatingly at America's industrial resources and population. Previous military technology had allowed America the luxury of waiting until wars became imminent, or had actually begun, before mobilizing. That luxury had vanished. The United States must prepare far in advance, since a modern war might be over in a matter of hours or days.

In short, the United States must embark on "a rapid and sustained build-up of the political, economic and military strength of the free world." The authors didn't cost out the buildup, although they did assert that the country could afford an increase of "several times present expenditures" on national defense and related foreign assistance. Devoting more than half the gross national product to such purposes, they contended, wouldn't break the American economy.[3]

Truman accepted NSC 68 in principle in April 1950, and he directed the various departments to work up funding estimates at once. Even a priority job like this took time, however, and while the sharp-pencil battalions labored into the sultry Washington summer, the Korean War began. Not surprisingly, the North Korean attack played directly into the hands of the Pentagon-expansionists. If anything, the attack suggested that the communists had advanced the day of reckoning.

Whether the subsequent tripling of the defense budget would have occurred in the absence of the Korean conflict is impossible to tell. For this reason, the objective impact of the exercise that produced NSC 68 can't be known. Yet the reductionist rhetoric of this remarkable document indicated the degree to which good-versus-evil thinking had penetrated the highest levels of the American policy apparatus. Acheson's after-the-fact remarks about bludgeoning the bureaucracy might suggest that those in the know understood the overstatements they were dealing in. Maybe they did. But words sneak up on people, even those who first utter them. Say something often enough, and you can easily come to believe it. A tendency to self-persuasion is an occupational requirement for politicians and salespeople generally. Certainly, Acheson's actions and those of others in the Truman administration yielded no evidence that they *dis*believed what they said.

Reds under the beds, or innocents abroad

If administration members lapsed into the language of Armageddon, they slipped understandably. The arena of American political debate during the early 1950s was slick with half-truths and smaller fractions. Joseph McCarthy greased the ground most egregiously with his swipes at the State Department, but the Wisconsin senator hardly lacked for company. Indeed, the telling phenomenon of the period wasn't the demagogues it produced. Every age in American history has had its slandering reputation-wreckers. What most marked the era of the early Cold War was the audience the demagogues attracted and the degree to which their unsophisticated arguments were supported by the subtler reasoning of some of the country's most acute minds.

For subtlety and fineness of distinction, not many matched philosopher and theologian Reinhold Niebuhr. And yet, while Niebuhr's anticommunism shunned the conspiratorial simplism of the McCarthyists, it fed the same impulses theirs did. In a 1953 essay entitled "Why Is Communism So Evil?," Niebuhr asserted that Americans and others were required to ask this question "because we are fated as a generation to live in the insecurity which this universal evil of communism creates for our world." He slighted those "timid spirits" who suggested that communism's opponents might be exaggerating the evil it embodied, and he dismissed observers who contended that traditional Russian imperialism, rather than communism, was what made Moscow dangerous. To attribute the Kremlin's crimes to Russian imperialism, he suggested, was to obscure the difference "between the comparatively ordinate and normal lust for power of a great traditional nation and the noxious demonry of this world-wide secular religion."

Niebuhr went on to judge communism "an organized evil which spreads terror and cruelty throughout the world and confronts us everywhere with faceless men who are immune to every form of moral and political suasion." Communism was more dangerous than Naziism, he said, because while the Nazis relied on force, the communists employed deception and treachery, posing "as the liberators of every class or nation which they intend to enslave." The fact that the communists claimed to be working toward a utopian goal of human equality merely heightened the danger they held for the West. Communist utopianism provided "a moral facade for the most unscrupulous political policy, giving the communist oligarch the moral warrant to suppress and sacrifice immediate values in the historical process for the sake of reaching so ideal a goal." The communists' conviction that after the emergence of the true communist state men would control their own destiny led them to usurp the powers of God. "This idea involves monstrous claims of both omnipotence and omniscience which support the actual monopoly of power and aggravate its evil." Hopes of some Americans that the incorporation of countries with different traditions, such as western-minded Czechoslovakia and Confucian China, into the communist realm might moderate Moscow's beliefs and behavior had proved vain. "Communism has been consistently totalitarian in every political and historical environment. Nothing modifies its evil display of tyranny."[4]

Though an intellectual ocean separated Niebuhr from McCarthy and the Wisconsin senator's imitators, Niebuhr's emphasis on the unmitigatedly evil nature of communism lent respectability to the McCarthyist onslaught. Niebuhr's warning about communism's penchant for duplicity reinforced the demagogues' insistence that restrictions on free speech and free thought were warranted if they helped protect America from the communist threat. By suggesting that the communists were usurping the authority of God, Niebuhr gave aid and comfort to those who wanted to launch a holy war against the Kremlin and all its works, or at least to employ tactics of the Spanish Inquisition against any who sympathized with the communist heresy.

Even without the backing of intellectual heavy-hitters like Niebuhr (who later expressed second thoughts about what he had helped unleash), the McCarthyists would have found plenty of encouragement. After all, some of their worries made perfectly good sense. Regarding the fear of spies in the defense establishment, for example, a reasonable argument could be made, and was, that no one had a constitutional or moral right to work for the Pentagon or in government scientfic laboratories. In the atomic age, the American government had to be extremely careful who gained access to secrets that might gravely jeopardize the physical security of the United States. (Although there was no single "atomic secret," there were some clever engineering tricks that if divulged might short-cut the Kremlin's path to parity with the United States.) It only required a few like Klaus Fuchs to put tens of millions of lives at risk, and simple prudence required indefatigable efforts to ferret out the Fuchses who continued to endanger America's existence.

A slightly more sophisticated argument could be used to justify the campaign against alleged subversives in the State Department. Diplomats didn't occupy quite such sensitive posts as atomic scientists, but an agent here or a dupe there in the councils of policy conceivably could tip America's hand to the Kremlin, or tip the balance of decision in favor of a course that would benefit the Russians and harm the United States. The fact that the potential for diplomatic damage to American security was less dramatic than in the case of atomic spies merely made the danger more insidious.

Justifying the campaign against communists and fellow-travelers elsewhere—in schools, in labor unions, in Hollywood—required a bit more reaching. It wasn't immediately apparent what the three Rs, collective bargaining, or the silver screen had to do with national security. Yet neither was it entirely inconceivable that leftist teachers could corrupt the nation's youth, that radical unionists could sabotage the nation's economy, or that pink scriptwriters could plant the seeds of defeatism in the minds of moviegoers. Once one accepted the Niebuhrian premise of communism as universal evil and communists as thoroughly unscrupulous, the distinction between caution and paranoia got fuzzy fast.

Of course, the anti-communist crusade also served purposes that had little to do with its professed fear for American security. Republicans red-baited Democrats out of simple partisanship. Democrats red-baited back, though less effectively. Populists scapegoated foreign-service officers, whom they disliked as bitterly for being too elitist as for being too liberal. Business leaders busted unions, or tried to, with anti-communist truncheons. (During the campaign for the Taft-Hartley Act of 1947, much was made of the measure's anti-communist provisions. In fact, its real target was the big unions.) Unreconstructed Roosevelt-despisers blasted the New Deal as kin to socialism. Truman's conservative opponents said the same about the Fair Deal. Segregationists called the president a communist for trying to integrate the armed forces.

Yet there was more than conscious self-interest behind the efforts to purge communism from American life, just as there was more than concern for American national security. The greatest appeal of anti-communism was probably the psychological security it provided Americans during a confusing and troubling time. Despite the fact that their country was the most powerful the world had ever seen—economically, politically, militarily— Americans felt themselves on the defensive. To be sure, American aid to

Greece had helped end the communist insurgency there (although Tito's decision to close his border to the rebels helped more). And the Marshall Plan was putting Western Europe back together again, thereby diminishing the credibility of communists in France, Italy, and elsewhere. The Berlin airlift had demonstrated that the West wouldn't be pushed out of that divided city. But still the Reds had gained ground. A 1948 coup had brought communists to power in Czechoslovakia. The communist victory in China had delivered the largest single portion of humanity to communist control. The North Korean attack on South Korea and the subsequent entrance of China into the fray showed the communists' willingness to complement subversion with military force in their efforts to extend their dominion. A mounting communist challenge to French control of Indochina indicated that the communist disease was spreading in Asia. Where it might crop up next could only be guessed.

Two explanations for communism's stunning successes presented themselves, one considerably less palatable than the other. The distasteful one held that communism was advancing on the strength of its merits. Communism, by this reasoning, really did have something to offer the downtrodden of the earth, just as Marx and Engels had said. (Marx and Engels hadn't had Chinese peasants and Vietnamese nationalists in mind, but to dwell on this was quibbling.) What had capitalism ever brought the peoples of Asia and Africa besides unequal treaties, colonialism, and repression? Communism promised independence, equality, and self-respect. Would the communists deliver on their promises? Only the future knew. But the present made plain the capitalists hadn't, when they had even bothered to promise anything beneficent.

While some liberals in the United States offered this explanation for communism's successes, it gained relatively few converts. The problem was that it required acceptance of ambiguity, of recognizing that good existed on both sides of the Cold War. Humans, as a group, commonly dislike ambiguity. Certainty is more satisfying. Americans historically have liked ambiguity even less than most peoples, largely because they haven't been required to live with it, at least not in relations with other countries. From 1783, Americans generally had their North American neighborhood to themselves. The indigenes caused a few problems, as did the British and Mexicans, but on the whole the northern half of the Western Hemisphere was America's oyster. While people in more crowded parts of the planet grew used to compromising with the folks across the border, Americans never got the hang of it.

People who have things their own way for a long time tend to develop exaggerated notions of their power and their rectitude. When no one keeps you from your desires, you feel powerful. When no one forces you to heed other viewpoints, you feel righteous. Both tendencies have beset Americans, and both contributed to widespread acceptance of the second explanation for communism's recent successes: conspiracy.

Conspiracy theories came naturally to those who wondered how communism continued to advance despite the opposition of the most powerful
country in the world. America wasn't losing the Cold War. It was being
sold out. Roosevelt and his fellow-traveling cohort—which included the
arch-subversive Alger Hiss—had handed Eastern Europe to Stalin at
Yalta. "Red Dean" Acheson and his cabal of State Department China
hands had delivered China to the Reds by abandoning Chiang. Truman
accepted stalemate in Korea rather than let General MacArthur fight.

Conspiracy theories were equally useful in allowing Americans to cling
to a belief in their own uprightness and innocence. There was a paradox
here, for if communism was succeeding because of the activities of collaborators in the United States, that would seem to imply American guilt,
rather than innocence. But the conspiracy argument could cut the other
way as well. Persons who abetted communist expansion had bartered their
souls to a foreign power. Regardless of the incidental fact of where they
were born, they were not *really* American. This argument worked better
for a nation of immigrants than it would have for most other nations.
Becoming an American was, or had been for ancestors, a matter of
choice—surviving American Indians and descendants of slaves not included. Therefore, *un*becoming an American was a matter of choice also.

The most important feature of conspiracy theories, and the feature that
makes them such perennial favorites, is their ability to divide the world
neatly into categories of good and evil. Humanity's problems aren't the
consequence of some abiding deficiency in all of us. Problems are the work
of bad people. *We* are not guilty. *They* are. Righting the wrongs of the
world requires no repentance from us, only from others.

During the early years of the Cold War, Americans needed to cling to
the myth of their innocence more than ever, for they were in the process of
repudiating the conditions that had fostered the myth in the first place.
The years after 1945 witnessed the obliteration of the Atlantic as a conceptual barrier between America and Europe. After the promulgation of the
Truman Doctrine, the launching of the Marshall Plan, and especially the
creation of the North Atlantic alliance, Americans could no longer think of
themselves as separate from the Old World—that portion of the globe they
had customarily deemed a cesspool of Talleyrandish intrigue and Bis-
marckian amorality. By the beginning of the 1950s, America was up to its
neck in the cesspool.

Eventually, Americans would discover that cesspools stain and infect.
But for the time being, a conspiracy theory of communist advance—
which was nothing more than a corollary to the two-worlds ideology of
American Cold War thought—provided a measure of prophylaxis. By
intellectually and psychologically concentrating evil in the regions con-
trolled by communism, Americans pushed America's moral frontier to the
Elbe, annexing Western Europe to the United States as part of a greater
America known as the "Free World." Despite forsaking the teachings of

the fathers, despite resorting to military alliances and other devices of traditional diplomacy, Americans could retain their exceptionalist notions and their belief in their country's innocence.

General Keynes

To protect their virtue, Americans armed to the teeth. Arguably, the most important effect of the Cold War on American life was the inspiration it provided for unprecedented spending on defense. Where Americans previously had waited until hostilities drew nigh before strapping on sword and buckler, during the early 1950s the country mobilized permanently, and spent accordingly. In 1950, the United States devoted just over $13 billion to defense, out of a gross national product of slightly under $285 billion, for a defense outlay of 4.6 percent of GNP. During the next few years, under the combined impact of NSC 68 and the Korean War, defense spending mushroomed, topping $50 billion in 1953 and constituting 13.8 percent of a GNP of $365 billion. Defense spending declined during the mid-1950s, as the fighting in Korea ended and the Eisenhower administration sought to contain federal expenditures generally. But it resumed its ascent during Eisenhower's second term, and in the last year of the decade the Pentagon and related agencies got $47 billion, or 9.6 percent of a $484 billion GNP.[5]

As with everything else about the Cold War, high defense spending served multiple purposes. The obvious purpose was strategic: to defend the United States, America's allies, and American interests around the world. By the thinking of NSC 68, the United States needed to devote a large portion of its gross national product to accomplishing one or both of two objectives: deterring the Soviets and their allies from attacking, and allowing the United States and America's allies to defeat the communists in case deterrence failed. NSC 68 was not the last word on the subject. It was more like the first. Throughout the 1950s (and for thirty years thereafter), the State Department, the Defense Department, the Central Intelligence Agency, assorted other government agencies, and a host of private think tanks updated threat assessments and devised new plans for countering advances in Soviet capabilities. All the thinking cost money, and most of it asserted that American security required spending more.

A second purpose of lavish defense spending was economic. During the Great Depression of the 1930s, officials of the Roosevelt administration had flirted with the notion that government expenditures might fill the shortfall in demand that was producing the bulk of the problems the country was experiencing. To some degree, the idea was simply common sense. If individuals aren't spending enough to keep the economy healthy, the government should do so. But common sense or not, it sounded dangerously radical. Orthodoxy required government to tighten its belt during hard times, along with everyone else. The idea of deficit spending gained a modicum of respectability through the work of John Maynard Keynes, whose 1936 *General Theory of Employment, Interest, and Money* placed the case for government intervention on a theoretically rigorous foundation. Yet it didn't gain sufficient respectability to succeed on its merits alone. Though Keynes earlier had won a following in the United States by his scathing denunciation of the 1919 Paris peace conference, many of these followers were precisely those conservatives who eyed bigger government with extreme suspicion. In any event, the improvising and atheoretical Roosevelt possessed neither the inclination nor the nerve to pursue a thorough-going policy of deficit spending.

He didn't until Pearl Harbor, that is. World War II added the clinching argument of national security to Keynes's more abstract reasoning. During the 1930s, the largest deficit Roosevelt could muster was $3.5 billion, in 1936. By 1943, however, the deficit had ballooned to a previously unthinkable $54 billion. The American economy responded vigorously to this injection of demand, just as Keynes had predicted. Between 1936 and 1943, the American GNP grew by nearly 75 percent in real terms. (In the previous seven-year period, from 1929 to 1936, GNP *fell* by more than 5 percent.)

In 1945, no one expected government spending to continue at anything like its wartime rate—which was the major reason for all the worries about the reconversion to peace, and for all the efforts to open up foreign markets and revive the European economy. Those efforts largely succeeded. Starting with the Bretton Woods compact, and continuing through the General Agreement on Tariffs and Trade, American negotiators working in tandem with like-minded representatives of other countries reduced barriers to international trade to their lowest levels in modern history. The Marshall Plan provided seed money the Western Europeans put to exemplary use. By the time the program liquidated itself in 1952, the Europeans had shaken their postwar paralysis and begun a period of unprecedented prosperity—which continued, with temporary and relatively modest setbacks, for forty years.

A prospering Europe and an open international environment for trade and investment might have allowed the American government to withdraw to prewar levels of participation in the economy without jeopardizing America's economic growth and good health. But the experiment was one Washington never attempted. American government spending dropped sharply from 1945 to 1948—during which period pent-up private demand worked to offset the drop—yet even in the latter year federal expenditures were more than twice what they had been in 1940, in real terms. From 1948 until the end of the 1950s, government spending nearly doubled again. Although Washington balanced its budget four times during the 1950s, during the other six years the bottom line ran red, and in 1959 the deficit once more hit double figures (of billions).

The principal contributor to the unprecedented growth in government spending was defense. Where the 1940 federal budget had devoted less than 16 percent of its total to defense, in 1959 defense claimed more than 50 percent. Put otherwise, if the government in 1940 had stopped payment on every defense purchase order, the direct impact on the American economy would have been a decline of between 1 and 2 percent—noticeable, but hardly devastating. The indirect, ripple effects would have amplified the consequences, yet the total damage would still have been modest. By contrast, a complete defense cancellation in 1959 would have dealt the economy a direct setback of nearly 10 percent, and the indirect effects, as laid-off tank-builders defaulted on mortgages and aircraft executives stopped taking Florida vacations, would have been hair-raising. Though no one conceived of eliminating defense spending entirely, anything approaching even a return to prewar levels could easily have triggered another depression. The point of national defense was to safeguard a way of life, not merely territory. And while pouring money into the Pentagon kept the Russian bear at bay overseas, it also kept the wolf of hard times from the door at home.

Brave New Look

The idea of a permanently large national defense establishment took some getting used to. The adaptable young military officers who discovered in procurement an avenue for career advancement accomplished the task, but Dwight Eisenhower, accustomed to the small-army atmosphere of the interwar period, never did. Moreover, Eisenhower as president led a party opposed to the bigger-government tendencies of the Democrats. As matters turned out, Eisenhower made no significant attempt to dismantle the New Deal, which in two decades of operation had spawned powerful constituencies. But his first inclination on examining budget figures was always to ask whether they might be reduced.

Six weeks after Eisenhower entered the White House, an event occurred that offered the possibility of substantial reductions in defense expenditures: Stalin died. The dictator's heirs, tremendously relieved at having outlived him, lost no time in proclaiming a new era in international relations. Where Stalin had asserted the incompatability of capitalism and communism and had proclaimed a fundamental antagonism between East and West, his successors declared that problems between the superpowers could yield to peaceful solutions, negotiated in a spirit of mutual understanding.

The Kremlin's new line struck a chord in much of the rest of the world. With fresh leadership in both Washington and Moscow, many people in various countries hoped for a moderation of the Cold War. The aging Winston Churchill, whose 1946 Iron Curtain speech had set the tone for the confrontations of the late 1940s, took a decidedly different approach now. Churchill advocated a summit meeting between the leaders of the

superpowers, at which the two sides might settle outstanding disputes in
Europe, Asia, and elsewhere. (Churchill included himself on his list of
invitees, which doubtless had something to do with his eagerness for a
meeting.)

Eisenhower rejected Churchill's call for a summit. The Republican
president had been elected on a platform denouncing the Democrats for
defeatism in the struggle against the communists. And while Truman had
pursued a policy of containment, which implied acceptance of communist
control of Eastern Europe, China, and North Korea, the Republicans
promised "liberation": the freeing of the captive nations from the Kremlin's grip. The Republicans—Eisenhower included—had especially castigated the Democrats for Roosevelt's role in the Yalta conference. When the
Republicans organized Congress at the beginning of 1953, a resolution
repudiating Yalta stood near the top of their agenda. As president, Eisenhower eventually came to reconsider the advisability of canceling the
agreement on which American rights in Germany, including Berlin, were
based. But so soon after the election, he could hardly repudiate his repudiation. The anti-Yalta resolution was working its way through committee
when the news of Stalin's death arrived.

Eisenhower had serious doubts regarding Moscow's sincerity in proposing to talk about peaceful resolution of conflicts. Doubts aside, he wasn't
willing to participate in anything that smacked of a repeat of Yalta. To
meet with the Soviets on a basis of equality would tend to confer legitimacy on communist hegemony in the territories the communists controlled. This would undermine Eisenhower's credibility, especially with
conservative Republicans whose support Eisenhower deemed crucial to
the effectiveness of his administration. Though Eisenhower personally
despised Joseph McCarthy, the president couldn't deny the Wisconsin
senator's effectiveness in softening up the Democratic opposition, and he
refused to criticize McCarthy in public.

Eisenhower didn't entirely rule out a meeting with Malenkov and who-
ever else might be in charge in Moscow. But he conditioned his participation on Soviet acceptance of terms that amounted to capitulation on key
issues between East and West. As he expected, these conditions killed
hopes for a summit.

If Eisenhower's refusal to talk sustained his political position among
American conservatives, it did nothing for his stature among American
moderates and liberals. It did still less for his and America's standing in
the world at large. Recognizing the risks of losing the peace issue to the
Kremlin, Eisenhower launched a propaganda counteroffensive. In April
1953, in a major address to the assembled panjandrums of the American
newspaper industry, he delineated the human costs of the ongoing superpower confrontation. "Every gun that is made, every warship launched,
every rocket fired, signifies, in the final sense, a theft from those who
hunger and are not fed, those who are cold and are not clothed." He went

on, "This world in arms is not spending money alone. It is spending the sweat of its laborers, the genius of its scientists, the hopes of its children. The cost of one modern heavy bomber is this: a modern brick school in more than thirty cities. It is two electric power plants, each serving a town of sixty thousand population. It is two fine, fully equipped hospitals. We pay for a single fighter plane with a half-million bushels of wheat. We pay for a single destroyer with new homes that could have housed more than eight thousand people." Eisenhower concluded, borrowing an image from fellow midwesterner William Jennings Bryan: "This is not a way of life at all, in any true sense. Under the cloud of threatening war, it is humanity hanging from a cross of iron."[6]

This speech, Eisenhower's best ever, received rave notices from the editors present. It had no effect on American Cold War policy, though. While the president titled his address "The Chance for Peace," he soon demonstrated his unwillingness to take any meaningful chances for peace. Eisenhower and other top American officials believed that the burden of proof in demonstrating good faith lay with the Soviets. Stalin's successors might speak more soothingly than the tough old Georgian, but until they demonstrated otherwise, by actions rather than words, the United States must operate on the assumption that they shared his world-conquering goals. If anything, their lulling statements merely made them more danger- ous. C. D. Jackson, a hard-liner who served as Eisenhower's special ad- viser for Cold War planning, remarked, "So long as the Soviets had a monopoly on covert subversion"—Jackson didn't consider American co- vert operations to be subversion—"and threats of military aggression, and we had a monopoly on Santa Claus, some kind of seesaw game could be played. But now the Soviets are muscling in on Santa Claus as well, which puts us in a terribly dangerous position."[7]

Eisenhower eventually would feel forced by the Kremlin's peace cam- paign to meet the Soviet leadership at Geneva in 1955. He also would decide to boost the Santa Claus side of American policy by increasing foreign aid. But for the time being, he concentrated on keeping America's powder dry. During the summer of 1953, Eisenhower ordered a re- examination of American national security policy. The exercise, labeled Project Solarium, was more elaborate than that which had produced NSC 68, and it had two objectives: first, to disavow, symbolically anyway, the doctrines of the defeated Democrats, and second, to devise a defense policy that would strike a balance between strategic and economic impera- tives. Eisenhower divided the Solariumites into three teams, which pro- duced three sets of policy recommendations. In all three reports, Moscow remained as malevolent as it had seemed to the NSC-68ers, but the three teams diverged regarding the precise nature of the Soviet threat and the appropriate American response. Team A advocated essentially an exten- sion of the containment policy of the Democrats. Headed, fittingly, by George Kennan, the A team argued that if the Free World held firm the communist challenge would diminish with time. "If we can build up and maintain the strength of the free world during a period of years," the Kennan group declared, "Soviet power will deteriorate or relatively decline to a point which no longer constitutes a threat to the security of the United States and to world peace." Looking to the day of communism's demise, American leaders should, without retreating to a policy of appeasement, respond cautiously to outbreaks of violence along the boundary of the Soviet sphere. The United States must be ready to fight a big war if the Russians insisted, but otherwise Washington should strive to keep local conflicts local.[8]

Team B advocated a more vigorously policed version of containment. Rather than preparing to respond to communist-inspired violence principally at the level and in the location where the violence took place, the United States should announce its willingness to resort to "general [that is, nuclear] war as the primary sanction against further Soviet Bloc aggression." Two advantages recommended this policy. First, the Soviets would think several times before provoking another Korea-type conflict. Second, by precluding Treasury-draining contests like Korea, the policy would save considerable money. "Since Alternative B rules out peripheral wars, its military costs will in the long term be less than the cost of any alternative that accepts such wars—by the amount those wars cost."[9]

Team C rejected containment on grounds that merely holding the line against communism wouldn't suffice to secure American interests. "Time has been working against us," the C team asserted. "This trend will continue unless it is arrested and reversed by positive action." The United States should exchange containment for liberation, aiming to raise the Iron Curtain and roll Soviet control back to traditional Russian frontiers. Military readiness would make up one aspect of such a policy, but only one. The United States must adopt "a forward and aggressive political strategy in all fields and by all means: military, economic, diplomatic, covert and propaganda." Such a policy wouldn't meet favor among some of America's allies. For this reason "the full scope of the plan would be revealed to them only gradually as successes were won." Nor would the plan be danger-free. It would entail "a substantial risk of general war." But better to fight and win than bleed and lose. "In this conflict one is either winning or losing," Team C concluded. "We are still losing."[10]

After letting the As, Bs, and Cs argue their cases for several weeks, Eisenhower attempted to combine what he deemed the strengths of the three positions. A congenital optimist and convinced democrat, he adopted A's dictum that time favored the Free World. A fiscal conservative and the leader of the historically tight-fisted party, he favored B's emphasis on relatively inexpensive nuclear weapons as the principal sanction against Soviet aggression. A career soldier and a personal witness of war's devastation, he opted for the diplomatic and covert aspects of C's anti-communist offensive, while frowning on that team's willingness to run the risk of general war. Liberation was a long-term project, not something to pursue precipitately.

In October 1953, Eisenhower approved an updated statement of American national security policy. This statement, labeled NSC 162/2, asserted that the death of Stalin had done little to alter the fundamental aims of Soviet foreign policy. "The basic Soviet objectives continue to be consolidation and expansion of their own sphere and the eventual domination of the non-communist world," the paper declared. Over time, though, the United States and its allies might look for a "slackening of revolutionary zeal" on the part of Soviet leadership, and for greater demands for consumer goods on the part of Soviet citizens. Together, these would render Moscow less manifestly dissatisfied with the international status quo.

This dulling of communist zeal would probably require years, as would significant changes in the global balance of power. "The detachment of any major European satellite from the Soviet bloc does not now appear feasible except by Soviet acquiescence or by war," the NSC paper asserted. The Soviets gave no indication of acquiescing in such detachment, and war would cost too much for the benefits it would produce. China, though not a Soviet satellite, acted as a loyal ally. "The alliance between the regimes of Communist China and the USSR is based on common ideology and current community of interests." Eventually differences might strain or rupture the alliance. "At present, however, it appears to be firmly established and adds strategic territory and vast reserves of military manpower to the Soviet bloc."

With neither defeat nor victory imminent in the Cold War, the NSC paper concluded, the United States must prepare for the long haul. It should develop and deploy capabilities for harassing the Kremlin. "The United States should take feasible political, economic, propaganda and covert measures designed to create and exploit troublesome problems for the USSR, impair Soviet relations with Communist China, complicate control in the satellites, and retard the growth of the military and economic potential of the Soviet bloc." More important, the United States must carefully balance military preparedness and economic vitality. Two sentences summarized Eisenhower's approach to national security policy: "A strong, healthy and expanding U.S. economy is essential to the security and stability of the free world. In the interest of both the United States and its allies, it is vital that the support of defense expenditures should not seriously impair the basic soundness of the U.S. economy by undermining incentives or by inflation." This solicitude for the health of the American economy, which most clearly distinguished Eisenhower's policies from Truman's, led to a decision to cut back on dollar-swallowing conventional forces and place greater reliance on comparatively cheap nuclear weapons. As it related to military planning, a single sentence encapsulated the new approach: "In the event of hostilities, the United States will consider nuclear weapons to be as available for use as other munitions."[11]

Less buck for the bang

This decision to conventionalize nuclear weapons, so to speak, formed the centerpiece of Eisenhower's "New Look" policy, and it gave rise to a nuclear strategy misleadingly known as "massive retaliation." The misleading began with a January 1954 speech by John Foster Dulles. Seeking, as ever, to distance himself and the Eisenhower administration from the policies of the Democrats, the secretary of state explained that the new policy would prevent a repeat of the Korean situation, where the Truman administration had allowed the United States to be suckered into an unwinnable war of attrition. To continue the Democrats' approach would require Americans "to be ready to fight in the Arctic and in the Tropics; in Asia, the Near East, and in Europe; by sea, by land, and by air; with old weapons and with new weapons." Preparing for such wars would produce "grave budgetary, economic and social consequences" for the country. The Eisenhower administration refused to travel this road. Instead, the president had made a basic decision "to depend primarily upon a great capacity to retaliate, instantly, by means and at places of our choosing."[12]

What Dulles seemed to be saying was that the United States would respond to another war like Korea with a threat or an attack against Moscow or Beijing. Such chest-beating made wonderful copy, but lousy strategy. While massive retaliation might have been conceivable in the pre-1949 days of America's nuclear monopoly, it seemed suicidally insane now that Moscow could respond to American nuclear threats and attacks with nuclear threats and attacks of its own. In August 1953, in the middle of the Solarium exercise, the Soviets tested a hydrogen bomb. What edge the United States had once enjoyed in the race for the biggest and most powerful weapons had now all but vanished. Would the West risk losing London or Paris or New York to save Seoul? Not if the British or French had a voice in the matter, and probably not if many Americans did.

Albert Einstein once described a principle that guided his search for laws explaining the operation of the universe. He aimed, he said, to make things as simple as possible, but no simpler. In dramatizing the difference between Eisenhower's policy on nuclear weapons and that of the Truman administration, Dulles overstepped the boundary between the simple and the simplistic. Robert Bowie, head of the State Department's policy planning board, later remarked that Dulles's desire to make a point sometimes led him into statements that didn't do justice to either Dulles or the issues involved. "In speaking," Bowie said, "he was so anxious to get things simple and clear and forceful, and to have them get attention, that he gave a picture of a mind which has all these qualities of simplification in black and white." Bowie thought the massive retaliation speech fell into this category. "I don't think that the speech was an adequate representation of his own thinking."[13]

Dulles subsequently attempted to clarify matters, but didn't succeed very well. He tried to explain how the new policy actually *increased* America's options. In the event of renewed aggression, massive retaliation against the Soviet Union or China would be one option, yet not the only one. Nuclear weapons might also be used on a local basis, as one might use somewhat oversized conventional forces. Then again, the administration might respond merely with conventional forces.

There were some serious problems with this position. By placing greater emphasis on nuclear forces, Eisenhower created a certain political momentum in favor of their use. During debates in 1945 over the possible employment of the atomic bomb against Japan, participants had pointed out that failure to use it would raise nettlesome questions. Politicians and taxpayers would wonder why the government had spent billions of dollars on a weapon Washington didn't have the guts to use. More damning would be the complaints of the relatives and friends of those who would die in fighting prolonged by a decision not to play America's trump card. Circumstances in the 1950s obviously differed from those in the last months of the war against Japan—the biggest difference being Moscow's ability to retaliate in kind—but some of the same influences applied. By committing publicly to reliance on the American nuclear arsenal, the Eisenhower administration made every confrontation with the communists a potential catalyst of nuclear war. To be sure, the possibility of escalation to the nuclear level (on one side at least) had existed since 1945, but by putting the case for nuclear weapons so directly, Dulles and Eisenhower shifted the center of gravity in the decision-making process. For Truman to opt in favor of nuclear weapons in the Cold War—as distinct from the war against Japan—would have required a positive decision to override an inclination toward conventional forces. For Truman, it was easier to say no to nuclear weapons than to say yes. For Eisenhower, the situation was reversed. Eisenhower could still say no. But he would have to work at it.

The work turned out to be hard. On several occasions, Eisenhower received strong recommendations to use nuclear weapons. During the 1954 Dienbienphu crisis—only weeks after the unveiling of the massive retaliation policy—the Pentagon put together a plan for a nuclear rescue mission, in which American bombers would save the besieged French from annihilation at the hands of the communist Viet Minh. Eisenhower, doubting that the rescue would be as clean and quick as its proponents suggested, said no.

Just months later, the pro-nuclearists were back. In the autumn of 1954, gun crews of the People's Republic of China began shelling positions held by Nationalist Chinese forces on the near-shore islands of Quemoy and Matsu. Again Eisenhower's military advisers, especially Arthur Radford, the chairman of the joint chiefs of staff, called for employing nuclear weapons against the attackers. Making the same argument that had surfaced in the deliberations over the Hiroshima bomb, Radford asked what America's nuclear weapons were for, if not for use in just such crises. Radford reminded the president that America's "whole military structure" rested on the assumption that the United States would use nuclear weapons in fights against the communists. To reject such use at the last minute would undermine America's credibility and throw American planning into confusion. Radford added that the United States would continue to have trouble with the Chinese until Beijing got a "bloody nose," and he recommended an unmistakable warning to the Chinese and the Soviets that the United States would use "all means available" to defend Quemoy and Matsu.[14]

John Foster Dulles shared Radford's fear that a refusal to follow through on the president's decision to allow the use of nuclear weapons would grievously constrain the administration. The secretary of state told Eisenhower that the administration would have to "face up to the question of whether its military program was or was not in fact designed to permit the use of atomic weapons." If, despite the communists' provocation, the American government refused to do what it had said it would do, who would believe American leaders in the future? Each crisis in which the United States didn't use nuclear weapons contributed to a popular feeling that it never would. This feeling would have a deleterious effect on American security. "We might wake up one day and discover that we were inhibited in the use of these weapons by a negative public opinion." Dulles shuddered at the thought. "Our entire military program would have to be revised." The secretary judged public desensitization to nuclear weapons to be a matter of "vital importance," and he said the administration should "urgently educate our own and world opinion as to the necessity for the tactical use of atomic weapons."[15]

This time, Eisenhower didn't say no. The president preferred to keep his options open. Fortunately for the peace of Asia, the Chinese chose to de-escalate the crisis before he had to decide one way or the other. Yet the outcome left those in the administration, such as Radford and Dulles, who believed in a psychological use-them-or-lose-them theory of nuclear strategy, as frustrated as ever.

There was another problem with the massive-retaliation policy, potentially more serious than the political and psychological one. This one was economic and bureaucratic: Nuclear weapons cost too little. Not everyone in the American defense establishment shared Eisenhower's frugal inclinations. On the contrary, after the adoption of NSC 68 and after the fat years of the Korean War, the Pentagon had no desire to return to the lean budgets of the early postwar period. The air force applauded the New Look, which called for enlargement and improvement of America's air-striking capacity, but the army nearly mutinied at the cuts the New Look dictated in ground forces. Army chief of staff Matthew Ridgway told a congressional committee that "the military power ratio between western defensive capability and the Soviet bloc's offensive capability is not changing to our advantage." The communist menace, Ridgway went on, could be seen "at every significant contact point between the Soviet bloc and the West." Yet despite this fact, "we are steadily reducing Army forces—a reduction through which our capabilities will be lessened while our responsibilities for meeting the continuing enemy threat remain unchanged." Ridgway acknowledged that new weapons, especially nuclear ones, might increase America's firepower. But he contended that the new weapons would only heighten the importance of ground troops. "Because of the increasing complexity of land warfare and the resultant greater battlefield demands upon the fighting man, the individual soldier, far from receding in importance, is emerging ever more clearly as the ultimate key to victory." The general concluded, "Man is the master of weapons and not their servant. He is the indispensable element necessary to achieve victory and will remain so in the foreseeable future."[16]

In this 1954 testimony, Ridgway didn't quite say that the Eisenhower administration had sacrificed military readiness to political imperatives. Ridgway was opposed to the new policy, but he wasn't overtly insubordinate. Yet when he retired in 1955—a retirement Eisenhower didn't try to defer—he made no bones about his suspicions. In a letter to Defense Secretary Charles Wilson, which soon became public, Ridgway enumerated America's military commitments and declared, "The present United States military forces are inadequate in strength and improperly proportioned to meet the above commitments, specific or implied." In his memoirs, rushed into print a short while later, Ridgway explained how he had argued against the budget cuts Eisenhower claimed were necessary to the health and strength of American society. He found the cutters' case incomprehensible. "There was no logic in this reasoning," Ridgway asserted. He added, "My bewilderment was increased by the fact that at the same time these reductions were being ordered in the Army's budget, economists, with the blessing of the administration, were hailing the country's greatest boom, predicting that within the next five years the national production was going to rise from $360 billion to $500 billion. If that were true, then I was not greatly impressed with the argument that $2 billion more in the Defense Department budget was going to bankrupt the country." With an almost audible "Aha!" Ridgway continued, "The real situation then dawned on me. This military budget was not based so much on military requirements, or on what the economy of the country could stand, as on political considerations."

Ridgway's views, of course, reflected no such narrow priorities, although in his next sentence he couldn't resist pointing out that "76 percent of the proposed reduction was to be made in Army funds." He stated, "We were in danger of again falling into that serious error that had placed us at such a grave disadvantage against an inferior foe in the first few months of the Korean War. We were subject to the same dangerous delusion, the misty hope that air power, armed with the fission or fusion bomb, could save us in time of trouble. To my mind this country could not adopt a more dangerous doctrine, nor one more likely to lead us down the path to war." This penny-wise pound-foolishness must stop.[17]

Say when

It would stop, and before long. The stopping would result largely from the complaints of Ridgway and those who thought similarly, but it would also result from an enormous expansion of American military commitments overseas. As Robert Taft had predicted, the North Atlantic pact of 1949 subsequently led to the arming of the Western Europeans and the garrisoning of Western Europe with American troops. The logic of the sequence of developments was not quite what Taft had envisioned, following more closely Lenin's argument that the key to relations between the West and Russia was China. The collapse of anti-communist forces in China, the outbreak of the war in Korea, and China's entry into the Korean fighting greatly alarmed American leaders, who feared that the Soviets would take advantage of America's entanglement in East Asia to assault Europe. These events equally alarmed European leaders, who feared that the turmoil in the Far East would provide ammunition to American isolationists and Asia-firsters, and perhaps provoke a pullout from Europe that would leave the British, French, Dutch, Belgians, Danes, and the rest to deal unaided with the Russians and the Germans (this latter threat only whispered, in polite company).

The coincidence of alarms occasioned the transformation of the Atlantic alliance into NATO—the transformation, that is, of a group of countries pledged to mutual defense in time of war to a multinational organization engaged in joint military activities and actively preparing for war during time of peace. It also occasioned the first part of the Western allies' answer to the German question, with the answer, in fact, playing a central role in the transformation. American and European strategists recognized that if war should break out in Europe, it would almost certainly begin in Germany. As a consequence, the defense of the western half of the continent should begin at the Elbe. Besides, the French and the other continentals in the alliance much preferred fighting in Germany to fighting on their home territories. Naturally, it would help matters if the Germans would assist in the fighting. And the Germans would be more inclined to fight if they felt they were fighting for their own country, which at present they did not have. Yet the idea of giving the Germans a country and supplying them arms unsettled lots of Europeans so soon after the recent war. A firm, physical American commitment would go far toward settling the skeptics.

This line of reasoning led in quick succession to the creation of the

Federal Republic of Germany, the approval by Congress of a $4 billion package of aid to speed the rearmament of Europe, the appointment of General Eisenhower to be supreme commander of allied forces in Europe (which enhanced Eisenhower's already good presidential prospects), the merging of the military organization of the narrow Western European Union into the broader North Atlantic Treaty Organization, and the assignment of American troops to NATO.

This last move, in particular, provoked Robert Taft and other doubters to call a halt. The legislative season of 1951 opened, in the Senate, with a "great debate" regarding the wisdom of stationing American troops in Europe. Truman was suggesting just a few divisions, but Taft spotted a camel's nose in the tent. "We have seen this process constantly repeated," he said. "We were told that Bretton Woods would solve the financial problems of Europe. Within a few months it became apparent that the British loan was necessary. That was followed by other aid, and then by the Marshall Plan." Each time, the administration had declared its measure vital to the morale and stability of Europe. Nor had the Marshall Plan, with all its billions, been the end. "Then came the Atlantic Pact, our obligation to go to war, and the commitment of small sums to the arming of European countries. Then the appropriations for arms had to be quadrupled. Now we are told that the morale of Europe will collapse unless we send at least three or four more divisions in 1951." Where would it cease? Taft predicted that in the absence of strict limits on American liability, the president would come back each year with requests for additional troops to safeguard the investment already sunk in Europe.

Taft criticized the administration's plan for a NATO army as not simply unnecessary but downright dangerous. "I believe that the formation of such an army, and its location in Germany along the Iron Curtain line, particularly one headed by an American general and dependent primarily on American strength, is bound to have an aggressive aspect to the Russians. We would be going a long way from home, and very close to the Russian border." Supporters of the administration might protest that the NATO force was merely defensive, but such protests would hardly reassure the Russians—with reason. "A first-class defensive army must also be an offensive army. Today, armies do not stand still; either one advances or the other advances." For this reason, the creation of the NATO force would threaten the peace and might provoke the Soviets to pre-emption. "The formation of this army is more an incitement to Russia to go to war rather than a deterrent."

The Ohio Republican didn't deny that a free Europe served American interests. Freedom everywhere served American interests. "It does not follow, however," he continued, "that because we desire the freedom of every country in the world we must send an American land army to that country to defend it." Let the Europeans defend themselves, with modest American help where essential. "Western Europe, after all, has more peo-

ple in it than we have in the United States, and is completely able to
defend itself if we furnish the armament and if it desires to do so."

Taft worried most that, in garrisoning Europe, the United States would
jeopardize its own defenses and undermine its own political institutions.
"If we commit ourselves to more than we can carry out, we weaken the
whole nation." The high taxes necessary to support the administration's
program might well ruin the country—as the president himself implicitly
admitted. "Even now, the president does not dare to recommend taxes to
meet the $71 million expenditure which he proposes. I do not see how we
can today raise taxes beyond, say, about $65 million a year without hard-
ship and injustice and danger to the economy of the country." On the
American economy, after all, rested the ultimate success of American
arms. Taft said he would gladly spend whatever American security genuinely required. But the president was carrying things too far and asking
too much. For the welfare of the country, the Senate must draw a line.
"We should not further endanger the position of America as the arsenal of
democracy and the bastion of liberty."[18]

Taft lost this round, as he had lost the previous ones. The American
troops went off to join the NATO army, which shortly added contingents
from new members Greece and Turkey to units from the original twelve.
Bringing the Germans aboard proved a bit more difficult, but in the
autumn of 1954 the fourteen offered to make the Federal Republic number
fifteen, and Bonn accepted.

Taft's fear that the Soviets might respond to the creation of a NATO
army by a pre-emptive attack turned out to be exaggerated, but Moscow
didn't intend to take the entry of most of Germany into the Western
alliance lying down. Nine days after the formalization of the Federal
Republic's membership in NATO, Soviet diplomats guided the hands of
the foreign ministers of the rump German Democratic Republic and the
other states of Eastern Europe as they put their signatures to the Warsaw
Pact.

From the American perspective, the extension and reinforcement of the
North Atlantic alliance formed but one part of an ambitious elaboration of
defensive responsibilities during the early and middle 1950s. In light of the
fact that events in Asia had triggered the transformation of the Atlantic
alliance into NATO, efforts by Washington to create mutual-defense agreements with the countries of non-communist Asia were only natural. Natural, but not straightforward: the creation of alliances with countries of the
southwestern Pacific, the Middle East, and Southeast Asia involved some
of the most roundabout reasoning and diplomacy American leaders ever
got mixed up in.

The American alliance with Australia and New Zealand had almost
nothing to do with American concern for the defense of those two countries. Not even the most Kremlin-phobic thinker in the State Department
or the Pentagon could devise a plausible scenario whereby the Soviets or

Chinese could challenge the safety or stability of either Australia or New Zealand. What Washington wanted from what became the ANZUS alliance was, first, Canberra's and Wellington's quiet acceptance of the revival of Japan, and second, Australian and New Zealand troops for use in the Middle East. The Australians, especially, and the New Zealanders, along with most of the rest of the peoples of the western Pacific region, still distrusted the Japanese deeply. While American troops occupied Japan, of course, the Japanese caused no trouble. But the 1951 Japan peace treaty restored sovereignty to the Japanese, and many who had felt the heel of Japanese warboots, directly or vicariously, worried whether the anti-militarism of MacArthur's Japanese constitution would stick long after the Americans left. The ANZUS treaty of 1951 offered reassurance that the United States still cared about the region. It also allowed Australia and New Zealand to plan to help Mother Britain in the Middle East in case trouble broke out there.

The Middle East wouldn't have been the Middle East without trouble, and American diplomacy toward that vexatious region was just as roundabout as toward Australasia. Washington first concentrated on Egypt, the most populous of the Arab states, the site of the Suez Canal, and the home of a large British base next to the canal. Unfortunately for Washington's anti-communist designs, the Egyptians sensed the British threat more easily than the Soviet threat. As Gamal Abdel Nasser, the prime mover of the military junta that ruled Egypt after a 1952 coup, told Dulles, "The Soviet Union is more than a thousand miles away and we've never had any trouble with them. They have never attacked us. They have never occupied our territory. They have never had a base here. But the British have been here for seventy years." Nasser said he would become a laughingstock in Egypt if he argued that the Russians posed a greater threat than the British. "Nobody would take me seriously if I forgot about the British."[19]

With Egypt uninterested in an anti-communist alliance, American attention shifted north, to the countries bordering the Soviet Union. Other considerations besides Nasser's veto contributed to the shift. In the first place, where better to draw the line against Soviet expansion than right at Russia's borders? This held particularly true regarding the oil fields of the Persian Gulf, which would be much easier to defend from positions in the mountains of Iran than from Egypt. In the second place, the Turks and the Pakistanis had long traditions of military prowess. A division of Anatolians from Turkey or Pathans from Pakistan would probably be worth two divisions of Egyptians. Third, although Muslims constituted large majorities in each of Turkey, Iran, and Pakistan, all three of those non-Arab countries—unlike Arab Egypt—cared relatively little about the plight of the Arab Muslims and Arab Christians displaced by the creation of Israel. Predominantly Arab Iraq, which would join the other three countries in the Baghdad Pact, felt the troubles of the Palestinian Arabs more directly, and in the end Iraq's hostility to Israel would help persuade Dulles and

Eisenhower not to seek American membership in the Baghdad organization. But for none of the four did the problem of Israel, which owed its existence in large part to American support, prevent association with the United States. As matters turned out, Turkey, Iraq, Iran, and Pakistan formally linked arms with each other and with Britain (London hoping to salvage something of its Middle Eastern empire), while the United States adopted the status of an informal but friendly observer. After Iraq dropped out following its 1958 revolution, America assumed formal membership in the renamed Central Treaty Organization, or CENTO.

By then, events elsewhere, notably in Southeast Asia, had made the "O" at the end of the acronym almost mandatory. When the Vietminh victory at Dienbienphu pulled half of Vietnam out of the Western orbit, Dulles organized yet another alliance to hold the other half. He ran into trouble getting Asian takers, and consequently the Southeast Asia Treaty Organization, or SEATO, had only two Southeast Asian members—Thailand and the Philippines—along with Pakistan (whose eastern half at least gave it frontage in the general area), Britain, France, Australia, New Zealand, and the United States.

In fact, for both SEATO and CENTO the ending "O" was an honorific signifying no substance. Only in Europe did the United States ever agree to the integrated command structure that made NATO a meaningful force-in-being. In the other regions, the Pentagon successfully resisted the resource-sharing a NATO-style organization required. Many American officials, not just in the Defense Department, recognized that countries like Pakistan had priorities other than anti-communism (anti-Indianism, in Pakistan's case) and that regimes such as Nuri-as-Said's in Iraq (and, for the more farsighted, Shah Pahlavi's in Iran) might fall to the forces of anti-Western radicalism. Better not to lean more heavily than necessary on such uncertain allies.

Another ally was more uncertain still. When the Chinese Communists ran Chiang Kai-shek off the mainland in the summer of 1949, few foreign observers expected the Nationalist leader to last long on Taiwan. At that point, the Truman administration would have been happy enough to see Beijing finish off the Nationalists at once. The China issue was already a disaster for the Democrats, and measured against the grand felony of losing China, misplacing Taiwan would count as no more than a misdemeanor. Unfortunately, the Korean War confused matters, and almost without thinking, Washington adopted the role of Chiang's guarantor. In the midst of the Taiwan Strait crisis of 1954–55, Eisenhower persuaded Chiang to withdraw from the least defensible of the offshore islands. But Chiang, among the hardest bargainers the American government ever had to deal with, demanded a mutual-security treaty in exchange. This treaty, concluded early in 1955, cemented the United States into the logically improbable, though politically quite plausible position of treating Chiang's regime on Taiwan as the sole legitimate government of China. Like certain other theories at serious variance with reality, the full consequences of this one would become apparent only with time, but for the moment it pleased the flat-earthers in Congress who thought the world ended just west of Quemoy and Matsu.

By the middle 1950s, the American alliance system girdled the globe. From North America it crossed the Atlantic via Greenland (courtesy of Denmark) and Iceland to Britain. It triangulated Europe between Norway's North Cape, Portugal's Cape St. Vincent, and Istanbul's Golden Horn. It traversed the highlands of Asia Minor, skirted the Caspian Sea through Iran and leaped the Hindu Kush to Pakistan and the Indian subcontinent. It turned the corner of Southeast Asia (with a branch angling down to the Antipodes), and ran up Asia's Pacific coast to South Korea and Japan before returning to North America.

Beyond this, the Western Hemisphere remained as fully a United States sphere as ever. Americas-watchers debated whether the Rio Pact of 1947, which nonbindingly proclaimed hemisperic solidarity against outside attack, or the 1948 charter of the Organization of American States, which unenforceably pledged the United States to nonintervention in the affairs of other members, meant the less. But no one seriously disputed the reality of the situation. In its own—generously self-defined—backyard, Washington would act as it wished, and the other countries would try to make the best of things.

What friends are for

The change from scarcely half a decade before was stunning. In the late 1940s, the American people had needed to be convinced that American interests required a commitment to defend the countries of Western Europe—countries that shared with the United States a common history, common culture, and common political and economic institutions. By the mid-1950s, Americans were pledged to the defense of one of the most polyglot and otherwise heterogeneous collections of territories in history. The American embrace included hereditary despots like Pahlavi in Iran, elected autocrats like South Korea's Rhee and Taiwan's Chiang, and a clutch of colonels in Pakistan, Thailand, and several Latin American countries.

About the only thing all the countries in the American system shared was an avowed opposition to communism. And in many cases the avowals seemed less than sincere, looking not far different from the professions of the "rice Christians" who filled the churches of Christian missionaries in pre-revolutionary China. Yet, as had been true in relations with Stalin during World War II, it was tempting and not difficult to overlook the shortcomings of allies. American leaders had taken the position that the crucial distinction in the modern world was between communists and non-communists, and if some American allies and clients practiced incomplete versions of democracy, at least Americans might hope that their country's good example would rub off. Rome wasn't built in a day, nor would the Free World be perfected in a decade. Meanwhile, in the life-or-death struggle with communism, the United States needed friends wherever it could find them. When the friends provided forward basing for American planes, ships, or troops, as many of the allies did, the United States needed them all the more.

The value of the allies transcended military strategy, however. For all America's power, Americans in the early 1950s felt beleaguered and beset, and though they devoted record amounts of their country's economic, political, and emotional resources to national security, they felt more insecure than ever. Their weakness for the arguments of the conspiratorialists was one measure of this national insecurity. As people in earlier ages had sought to exorcise evil by pinning the blame on goats and demons and witches, Americans blamed communists. Another sign of national insecurity was the desire, becoming by the mid-1950s almost an obsession, for allies. Whether or not all the allies and clients made America more secure militarily—and some, like South Vietnam, seemed military liabilities rather than assets—they made Americans feel better about themselves and the course they had chosen for their country. Just as individuals get a shot of confidence from knowing that other individuals share their beliefs and values, so countries take comfort from discovering like-minded countries. This is especially true for countries that hold up their value systems and political institutions as examples for others. It is truer still for a country that feels itself under ideological attack from a country with a competing vision, as America did during the 1950s.

In such an atmosphere, every country that could be enlisted on the side of the United States in the Cold War diminished Americans' sense of insecurity and reinforced their self-confidence. If the enlistees, in fact, had little in common with the United States, and joined forces with America chiefly for the military and economic aid an American alliance entailed, they generally had sense enough to say the words Americans wanted to hear.

American officials, in their turn, also played the game. Washington could be counted on to praise allies and clients for their opposition to godless communism, if not for their strict observance of the human rights and civil liberties of all their subjects. It was convenient, and not completely coincidental, that three of America's principal protégés in Asia—Chiang, Rhee, and South Vietnam's Ngo Dinh Diem—were Christians.

The flip side of the situation was that each country that was lost to communism threatened America's self-esteem. If the country succumbed to Soviet military force, as was the case with most of Eastern Europe, the succumbing signaled a shortfall of American strength. This was bad, but not as bad—for American self-esteem—as if the country went communist uncoerced by Soviet power. The thought that people might choose communism on their own was sorely troubling, for it indicated a rejection of the American model, and thereby suggested that American values were not what they were cracked up to be. The outcome of the Chinese revolution had been so traumatic for Americans for precisely this reason. The Chinese were rejecting America and American values. American officials and other keepers of the public conscience attempted to mitigate the feeling of rejection by asserting that Mao and his cadres were stooges of Moscow. But the experience seared the American psyche for a generation. Preventing another China, with all the injury another China would inflict, was a primary objective of the alliance-building of the 1950s.

Yet there loomed a scenario even more traumatizing than that involving China. In the Chinese case, the communists assumed power by force of arms, and consequently it was possible to describe their victory in the conspiracy terms that appealed to Americans for other reasons. What would have been utterly devastating was for communists to take over a country by peaceful, parliamentary means. A conspiracy that comprised a working majority of an electorate hardly counted as a conspiracy.

In two cases during the early 1950s, Washington went to great lengths to prevent such peaceful takeovers. In 1953, the Central Intelligence Agency helped overthrow the elected government of Mohammed Mossadeq in Iran, who seemed to American officials overly dependent on the communist Tudeh party. The overthrow was either a coup or a countercoup, depending on one's reading of the Iranian constitution, and it was by no means an exclusively American operation. British agents had a hand in the affair, as did plenty of Iranians who disliked and distrusted Mossadeq. Mossadeq might even have fallen of his own accord. Yet in the sequence of events that actually did take place, American money and American assurances of friendliness toward a post-Mossadeq government proved crucial in tipping the scales against the prime minister.

In Guatemala in 1954, the United States played a larger role. This time, the CIA trained and outfitted a small army of Guatemalans opposed to the elected reformist president, Jacobo Arbenz Guzman. As with Iran and Mossadeq, American officials feared that Arbenz, while not himself a communist, was paving the road that would bring communists to power. When the CIA-organized insurgency's ground attack on the government stalled at the decisive moment, Eisenhower authorized the spooks to take to the skies. Planes piloted by American agents bombed Guatemala City, effecting little physical damage but convincing Arbenz that Washington was committed to his ouster. Arbenz fled, leaving control of the country to a murderously pro-American general, Carlos Castillo Armas.

In each of the Iranian and Guatemalan cases, the American government presented the restoration of anti-leftist forces as a vindication of the values shared by the countries of the Free World. (Allen Dulles, the director of the CIA, couldn't resist some not-for-attribution bragging, but officially Washington had nothing to do with either affair.) In each case, the restora-

tion served strategic purposes. In Iran, it held the line against a possible
Soviet advance southward, while in Guatemala it prevented the potential
establishment of a pro-Soviet government just north of the Panama Canal.
In each case, it served economic objectives. The Iran operation ended with
the grateful shah's granting American oil companies a stake in what previously had been a British monopoly, while the new government of Guatemala rescinded an Arbenz-sponsored measure nationalizing nearly a
quarter-million acres of land titled to the American-owned United Fruit
Company. Yet the psychological element was no less important for being
reinforced strategically and economically. In the American view, the three
factors clustered, and right-thinking governments abroad—governments
that supported American values, by words if not always by deeds—would
naturally do right by American strategic and economic interests.

The events in Iran and Guatemala showed Americans reacting as imperialists have nearly always reacted to challenges to their hegemony. Most
Americans, of course, roundly rejected any intimation that their sphere of
influence was an empire. Not for nothing was it called the Free World.
And indeed the American sphere didn't much resemble the directly ruled
empires of Britain and France. But if an essential characteristic of empire is
control—the ability to control the activities of the various peoples within
the empire—then, in important respects, the American sphere functioned
as an empire, regardless of what Americans preferred to label it. The
manipulation of Iran and Guatemala demonstrated that the United States
could, and would, whack down regimes that attempted to break loose
from American control. Although American influence with the countries
of Western Europe normally fell short of control, on such occasions as the
Suez crisis of 1956 Washington abruptly yanked the British and French
into line by irresistible economic and political sanctions. In 1958, Eisenhower ordered American troops into Lebanon to prevent that country
from slipping out of the American sphere. During most of the 1960s and
the first part of the 1970s, the United States would wage what amounted
to the largest imperial war of the century to forestall the loss of South
Vietnam to the communists.

Empires or spheres of influence—call them what you will—generate a
momentum of their own. Possession confers a variety of benefits on the
possessors, not the least being a feeling of stature and self-confidence. Lose
the empire, and the feeling disappears. Or, at any rate, it is compelled to
go in search of a new justification. For this reason, if for no others, the
possessors resist the loss. Holding what one has consequently assumes
great significance, for Americans as much as for other imperialists.

★

CHAPTER THREE

The Immoral Equivalent of War
1955–1962

Moscow rules

The CIA operations in Iran and Guatemala represented only a small part of American covert activities during the first phase of the Cold War. The intelligence agency had grown out of the Office of Strategic Services of World War II, although not directly or immediately. As the world war approached its end, the director of the OSS, William Donovan, urged Roosevelt to keep the intelligence service together. The United States, Donovan predicted, would encounter problems of intelligence-gathering and analysis in the postwar period no less formidable than those it was encountering during the war. "We have now in the Government the trained and specialized personnel needed for the task," Donovan said. "This talent should not be dispersed."[1]

Whether or not Roosevelt would have dispersed the spies, Truman did. Roosevelt's successor realized he would have a full plate controlling the government agencies he could see, and he had no desire to add to his problems by encouraging bureaucrats who worked out of sight. But the deepening troubles with the Soviet Union forced Truman to change his mind, and in 1947 he put his signature to a measure that in one swoop created the Department of Defense (out of the War and Navy departments), the National Security Council, and the CIA.

Initially, the intelligence agency was just that: an agency for the collection and evaluation of intelligence regarding the security of the United States. Its authorizing statute gave the CIA no power to conduct covert operations. The framers of the legislation considered covert operations an aspect of war. If war came, the army, navy, and air force would handle the situation.

Yet Congress had thoughtfully—or thoughtlessly, depending on the members' attitudes toward covert operations—included an elastic clause in the agency's charter, authorizing the CIA "to perform such other functions and duties related to intelligence affecting the national security as the National Security Council may from time to time direct." Almost before the charter came back from the printer, the National Security Council decided to direct such additional duties. The confrontation between the superpowers had continued to sharpen during the first half of 1948, prompting Truman and the NSC to approve a top-secret paper calling for the creation of an office of "special projects." The paper's preamble justified the new office as a response to "the vicious covert activities of the USSR, its satellite countries and Communist groups to discredit and defeat the aims and activities of the United States and other Western powers." The paper went on to define acceptable American covert operations as those "conducted or sponsored by this Government against hostile foreign states or groups or in support of friendly foreign states or groups but which are so planned and executed that any US Government responsibility for them is not evident to unauthorized persons and that if uncovered the US Government can plausibly disclaim any responsibility for them." Getting down to specifics, the paper asserted, "Such operations shall include any covert activities related to: propaganda, economic warfare; preventive direct action, including sabotage, anti-sabotage, demolition and evacuation measures; subversion against hostile states, including assistance to underground resistance movements, guerrillas and refugee liberation groups, and support of indigenous anti-communist elements in threatened countries of the free world."[2]

While Truman remained president, the covert operatives had to content themselves with modestly intrusive activities like channeling secret funds to anti-communist political parties in Western Europe. Truman never lost his distrust for agents engaged in activities on the underside of American diplomacy. Without a firm mandate from the White House, the CIA stayed largely clear of matters of high policy.

The election of Eisenhower changed things dramatically. Eisenhower had discovered the possibilities inherent in covert operations during World War II, when the OSS and the military's own psychological warfare teams had worked in conjunction with resistance groups in German-occupied territory to distract the defense and prepare the way for Eisenhower's invasion forces. Further, after Korea had revealed the weaknesses of limited warfare, and as the development of a Soviet nuclear arsenal demonstrated the folly of general conflict, Eisenhower conceived of covert operations as an essential element of America's Cold War–waging capabilities. As evidence of his enthusiasm for covert operations, Eisenhower appointed Allen Dulles, director of CIA covert operations during the last years of the Truman administration, to head the entire American intelligence community.

The appointment occasioned a few problems, not the least being Dulles's aversion to anonymity. Dulles blamed the nature of American society for his agency's inability to keep hidden what he thought needed to be hidden. "England has the tradition for it and is able to operate in some secrecy," he told a group of reporters—who refrained from remarking that the head of Britain's MI6 didn't make a habit of meeting with the press. Dulles lamented, "That is not possible in this country."[3]

To stem the leaks that regularly emanated from Dulles's agency, Eisenhower enlisted General James Doolittle, the hero of the first air raid over Tokyo. The president directed Doolittle to conduct a review of the activities of the CIA and other agencies engaged in covert operations, and to recommend steps to improve their performance. Doolittle responded with specific comments about the managerial style of Allen Dulles and broader reflections on the mission of American intelligence. While conceding that the CIA probably required "a strange type of genius" to run it, and that Dulles might approximate the necessary qualifications better than anyone else available, Doolittle complained that entirely too much information was being served with the drinks on the Georgetown cocktail circuit. The president, he said, must do something to stop the talking.[4]

Doolittle's broader remarks were the more important. The general thoroughly endorsed American Cold War orthodoxy. "It is now clear that we are facing an implacable enemy whose avowed objective is world domination by whatever means and at whatever cost," he declared. From this sobering but by now pedestrian premise, Doolittle proceeded to a logical but nonetheless chilling conclusion. "There are no rules in such a game. Hitherto acceptable norms of human conduct do not apply. If the United States is to survive, long-standing American concepts of 'fair play' must be reconsidered. We must develop effective espionage and counter-espionage services and must learn to subvert, sabotage and destroy our enemies by more clever, more sophisticated, and more effective methods than those used against us. It may become necessary that the American people be made acquainted with, understand and support this fundamentally repugnant philosophy."[5]

Doolittle had seen flak over Tokyo, but the worst the Japanese gunners had sent this way would have been nothing next to the outrage produced by an admission that the United States engaged in the sort of nastiness the Soviets did. Needless to say, Eisenhower had no intention of declaring that the United States had abandoned its claim to moral superiority in the Cold War and henceforth would operate under the same rules of bleak efficacy that governed Moscow's behavior.

Yet if Eisenhower rejected Doolittle's public-relations advice, the president accepted Doolittle's substantive recommendations. Flushed with the success of the CIA operations in Iran and Guatemala, Eisenhower gave his covert warriors almost carte blanche to "subvert, sabotage and destroy" America's enemies. In some instances, the subversion was relatively innocuous. American agents launched balloons from Western Europe loaded with small consumer items like disposable razor blades, the purpose being to engender dissatisfaction at the meager living standards downwind of the

Iron Curtain. American propagandists, and European broadcasters funded by American money, beamed programs into Eastern Europe, hoping to convey a conviction of the duplicity of the communist regimes and the achievements of the Western democracies. Sometimes this psychological warfare inadvertently crossed over into the real thing. Broadcasts from the West evidently encouraged the Hungarian rebels of 1956, who died by the thousands when concrete Western support failed to materialize.

Other schemes were more deliberately ambitious. American agents endeavored to consolidate the position of Ngo Dinh Diem in South Vietnam and to unconsolidate Ho Chi Minh's in North Vietnam. American money bought votes in Syrian elections. American advisers assisted illegally in the election of Philippine president Ramon Magsaysay, and kept in close touch with Magsaysay after the election. American operatives provided financial and logistical support for an uprising against the Sukarno government of Indonesia.

The most objectionable schemes, however, involved attempts to assassinate foreign radicals. One operation targeted Fidel Castro, who found himself at odds with Washington shortly after seizing power in Cuba in 1959. The two sides first traded shots in an economic war, with Castro expropriating American holdings in Cuba, and Eisenhower canceling Cuba's access to the American sugar market and embargoing most exports to the island. When Castro turned to the Soviet Union for aid and comfort, American officials decided to take more-drastic measures.

At the end of 1959, Allen Dulles received a memo from J. C. King, the CIA's top official for Latin America, recommending that "thorough consideration be given to the elimination of Fidel Castro." King contended that the Cuban revolution and Castro were nearly synonymous. Informed sources in Cuba, King said, indicated that "the disappearance of Fidel would greatly accelerate the fall of the present government."[6]

Dulles directed King to look into means of "eliminating" Castro. During the early months of 1960, King did so. His investigations revealed some problems with assassination. As American analysts got a better look at the new regime in Havana, they came to suspect that killing Fidel would not undo the revolution. Fidel's brother Raul and the Argentine expatriate Ernesto (Che) Guevara might push the government in an even more radical direction. At a March 1960 meeting, King told the CIA task force in charge of policy toward Cuba that assassination didn't appear promising. "Unless Fidel and Raul Castro and Che Guevara could be eliminated in one package—which is highly unlikely—this operation can be a long, drawn-out affair, and the present government will only be overthrown by the use of force."[7]

American planning against Castro thereupon shifted toward just such use of force. A week after this meeting, Eisenhower signed off on a proposal to train Cuban emigres for an invasion of Cuba.

But the designs on Castro's life didn't die. Contingency assassination plans continued. During the summer of 1960, the CIA contacted elements of the American Mafia about a contract on Castro's life. For awhile, nothing came of the contacts. Priority went to the invasion by the emigre army. Yet the agency would pull the assassination schemes off the shelf after the invasion, at the Bay of Pigs, failed.

The United States worked harder during this period to kill Patrice Lumumba, the charismatic Congolese leader. Lumumba antagonized Washington amid a most confusing devolution of power from Belgium to the new government of the Congo (later Zaire). Belgian officials pulled out in June 1960, leaving a crossfire among various groups struggling for the succession. The tens of thousands of Belgian nationals remaining in the country particularly felt the crossfire, and within weeks of leaving, Belgium returned, in the persons of paratroopers sent to secure the safety of the left-behinds. Prime Minister Lumumba responded to this invasion by calling, first, on the United Nations, and then on the Soviet Union, for help in throwing out the invaders.

The Eisenhower administration couldn't much complain about an appeal to the United Nations, but it disliked immensely Lumumba's turning to the Soviet Union. Eisenhower declared that the situation was reducing to "one man forcing us out of the Congo," the one man being "Lumumba supported by the Soviets." A short while later, Eisenhower's national security adviser, Gordon Gray, went from a conference with the president to a meeting of the administration's special committee for covert operations to say that Eisenhower had expressed "extremely strong feelings on the necessity for very straightforward action" to remove this turbulent fellow. The president had examined plans for undermining Lumumba politically and had indicated doubt that the plans were sufficient. When Gray and the special committee recommended that the CIA take measures against Lumumba, they pointedly declined to preclude "any particular kind of activity which might contribute to getting rid of Lumumba."[8]

The following day, Allen Dulles cabled the CIA station in Leopoldville, informing officers there of the "clear-cut conclusion" in Washington that Lumumba couldn't remain in power without threatening grave harm to the United States. "The inevitable result will at best be chaos and at worst pave the way to Communist takeover of the Congo with disastrous consequences for the prestige of the UN and for the interests of the Free World generally." Lumumba must go, Dulles said. "We conclude that his removal must be an urgent and prime objective and that under existing conditions this should be a high priority of our covert action." Dulles authorized the CIA station chief to employ "more aggressive action" against Lumumba than heretofore, "if it can remain covert." The station chief should use his own discretion as to particular means, and shouldn't feel constrained to consult back with Washington. "We realize that targets of opportunity may present themselves to you."[9]

During the next few weeks, CIA operatives in the Congo met some of

Lumumba's Congolese opponents who were hatching plots of their own. By now, Lumumba had been forced from office and found himself surrounded by enemies. The station chief, Victor Hedgman, encouraged the plotters, urging the arrest or "more permanent disposal" of Lumumba.[10]

At the same time, CIA officials in Washington devised techniques for doing away with Lumumba in case the Congolese conspirators muffed the job. Richard Bissell directed the agency's biological warfare bureau to select a pathogen to knock Lumumba "out of action." The germ-warriors complied, and Washington shipped the bug to Leopoldville. Agents there tried to figure out how to expose Lumumba to the pathogen—to put it in his food, on his toothbrush, or on anything else that might go into his mouth. Not surprisingly, Lumumba exercised great care in such matters, and the microbe scheme came to nothing.[11]

As a backup plan, Hedgman suggested simply shooting Lumumba. This proved no easier, since Lumumba didn't venture out much. Hedgman placed a watch on Lumumba's residence, and had to report to Washington in November: "Target has not left building in several weeks."[12]

Days after Hedgman sent this cable, Lumumba did leave the building. On a dark and rainy night, he slipped past his watchers and headed for Stanleyville and, he hoped, safety. On the way, his Congolese rivals caught up with him. Sometime early in 1961, they killed him in Elisabethville. American agents applauded the result, but apparently had no direct hand in it. On learning of Lumumba's death, the station officer in Elisabethville wired Washington, "Thanks for Patrice. If we had known he was coming we would have baked a snake."[13]

Dr. Strangelove, I presume?

Researchers studying human emotions have noted a feedback mechanism between the outward manifestations of emotions and the internal perceptions of those emotions. I cry because I feel sad, to be sure. But, to some degree as well, I feel sad because I cry. Or, seeing myself crying, I know I ought to feel sad, and do. An analogous form of feedback—the Sotheby's syndrome—can be found in people who spend lots of money on singular items whose objective value is difficult to determine. Having paid lots of money, they proceed to convince themselves that the items merited the expenditures.

Something similar connected American actions in the Cold War, especially relating to assassinations and kindred crimes, to American perceptions of the communist threat. American leaders were, for the most part, morally aware and responsible individuals. Yet they ordered attempts on the lives of foreigners like Castro and Lumumba, who had done nothing more heinous than pursue political agendas unfavorable to certain American interests. How did Eisenhower and Kennedy and others involved justify their actions? They did so by claiming that the communist threat was so great as to justify almost any countermeasure. In other words, they tried to kill Castro and Lumumba because Castro and Lumumba appeared to be playing into communist hands; but the very fact of their trying to commit murder reinforced their perceptions of the direness of the communist danger. Though they might not have articulated the reasoning, it probably went something like this: We are good people working on behalf of a just cause. We are attempting to kill Castro and Lumumba. We wouldn't be doing so unless they represented a critical threat to American security. Therefore, they must represent such a danger. As in so much else related to the Cold War, the consequence became the cause, in a mutually reinforcing cycle.

To the degree that such a mechanism affected American policy during the early 1960s, it affected only those few policy-makers aware of the assassination plots. Public knowledge would have to wait until a later decade. In another area, though, the feedback between consequence and cause operated in the open much sooner. During the latter half of the 1950s, such rationale as had supported the doctrine of massive retaliation broke down entirely. Reports by the CIA, by the Pentagon, and by committees of outside consultants indicated that the Soviet Union was rapidly acquiring the ability to annihilate America. Offensive military technologies—especially medium- and long-range missiles armed with nuclear warheads—were far outstripping technologies for defense, laying the United States open to unprecedented physical devastation. In 1955, a committee headed by James Killian of the Massachusetts Institute of Technology reported the possibility of "death and destruction on a scale almost beyond knowing and certainly beyond any sensibility to shock and horror that men have so far experienced." The Killian report went on to declare, "For the first time in history, a striking force could have such power that the first battle could be the final battle, the first punch a knockout." Six months later, the Atomic Energy Commission described recent developments in warhead-miniaturization and rocketry that would soon make intercontinental nuclear-tipped missiles a reality. The commission's John von Neumann explained to a meeting of Eisenhower's National Security Council that Soviet deployment of ICBMs would allow the United States only fifteen minutes' warning in case of attack. Von Neumann added that the missiles were "horribly difficult" to intercept and, being relatively cheap (about $1 million each), could be produced in large numbers. Meanwhile, the Pentagon was compiling a separate assessment noting that the Soviets could easily conceal nuclear weapons in the holds of cargo ships, and lay waste major American cities from harborside.[14]

The news grew worse with passing time. In October 1957, Moscow placed the first mechanical satellite in orbit. American observers required only a rudimentary knowledge of physics, which was readily supplied by helpful editorialists, to guess that rockets that could send a Sputnik clear around the world and into orbit could send warheads less than halfway

around the world and onto American soil. The alarm over Sputnik touched off congressional investigations—energized by Democrats eager to
find a chink in the armor of the ever-popular Ike—into why and how the
Soviets had managed to surpass the United States technologically.

The Sputnik furor also lent weight to another report, already in the
works, on the perils of the nuclear age. When the Gaither report was
finally declassified in the 1970s, William Proxmire described it as one of
the handful of decisive documents in the development of American Cold
War strategy. (Proxmire placed the Gaither report next to George Kennan's 1947 "X" article defining containment, and NSC 68.) The Wisconsin senator may have been right, although Eisenhower did his best to
prevent the Gaither paper from gaining such stature. Eisenhower failed,
not least because of the stature of the individuals composing the Gaither
panel—for which Eisenhower had only himself to blame, since the White
House had organized the panel. H. Rowan Gaither, Jr., of the Ford Foundation headed an advisory group that included General Doolittle, now of
Shell Oil; Admiral Robert Carney, formerly chief of naval operations and
currently with Westinghouse Electric; Ernest Lawrence of the University
of California's radiation laboratory; Mervin Kelly of Bell Telephone's research division; banker-diplomats Robert Lovett and John McCloy; and
Frank Stanton of Columbia Broadcasting System. Paul Nitze, the principal author of NSC 68, helped draft the report.

Such a solidly established group couldn't be expected to challenge the
general philosophy of American foreign policy, and it didn't. "We have
found no evidence in Russian foreign and military policy since 1945," the
authors declared, "to refute the conclusion that USSR intentions are expansionist and that her great efforts to build military power go beyond any
concepts of Soviet defense." Yet if the nature of the Soviet threat hadn't
changed, its severity had. The Soviet economy, though still less than half
the size of the American, was growing much faster. Moreover, the Soviets
were devoting a much larger proportion of their domestic production to
military uses. Using striking graphics to bolster their case, the Gaither
authors noted that while Americans produced 58 times as many automo-
biles as the Russians, and 49 times as many clothes-washers, Soviet
machine-tool production doubled that of the United States. Russia's momentum apparently was inexorable. Another visual display showed the
trend lines of Soviet and American "military effort." The American line
for the past several years was above the Soviet line. The lines crossed at a
point corresponding to the present. The line for the Soviet future rose well
above the American line. At current rates, the Soviet lead could become
"critical" by 1959 or 1960.

As a consequence, the United States had no choice but to increase its
military outlays significantly. It must enhance the survivability of the
American nuclear force by placing more of the B-52 fleet on alert (which
wouldn't cost much), by improving radar and other early-warning devices

(which would cost more), and by developing an active missile defense system (which would cost a lot). The government should also commence construction of fallout shelters for the American civilian population. Regarding this last item, the Gaither report rejected scenarios that showed a nuclear attack to be so horrific that no one would survive or want to. "We are convinced that with proper planning the post-attack environment can permit people to come out of the shelters and survive."

Safety came dear. The full Gaither program would add nearly $5 billion to the administration's projected federal budget of $38 billion for fiscal 1959, and would rise to nearly $12 billion extra by 1961. Yet America could carry the burden. "The nation has the resources, the productive capacity, and the enterprise to outdistance the USSR in production and in defense capability." During World War II, the American people had paid over 40 percent of the gross national product for defense. The expenditures required to meet current conditions would amount to scarcely a third of that.

Popular opinion held the key to making the required effort possible, the Gaitherists said. "The American people have always been ready to shoulder heavy costs for defense when convinced of their necessity." The government must take the lead in the convincing. The report urged "an improved and expanded program for educating the public in current national defense problems." The effort couldn't come too soon. "We must act now to protect, for this and succeeding generations, not only our human and material resources but our free institutions as well."[15]

Eisenhower attempted to suppress the Gaither report, judging it alarmist and short-sighted. But the main points leaked to the press, and added to the uproar created by Sputnik. For the first time in their history, Americans found themselves facing the specter of national extinction. Psychologically, it helped only a little—though strategically it meant the world, literally—that the Soviet Union shared their predicament. Few peoples have placed such store in the future as Americans, for whom each generation promised, and usually delivered, greater wealth and power than the previous generation. Odd dyspeptics like the American Brothers Grim— Henry and Brooks Adams—read decline into the American experience, and setbacks like the depression of the 1930s caused more than a few to wonder whether American progress had in fact run its course. But for most Americans through most of American history, the future promised only more and better. Now Americans had to deal with the prospect that the future might hold nothing at all. Modern technology might write a conclusion to the American experiment in the wink of an eye.

Or the finish might take longer. A concomitant fear to that regarding a nuclear war involved the steady poisoning of the atmosphere, the oceans, and the food chain by radioactive fallout from weapons testing. The deadly character of testing had become apparent in 1954 when a Japanese fishing vessel, ironically named *Lucky Dragon*, suffered the misfortune of

trawling leeward of an American hydrogen-bomb test in the Pacific. The
crew quickly fell ill, and one man died. American apologies and repara-
tions couldn't silence worries that the accident foretold the fate of millions
of other people. During the next several years, scientific studies showed
the accumulation of such toxins as radioactive strontium in water, soil,
agricultural produce, and even mothers' milk. Medical specialists predicted a rise in the incidence of cancer, while biophysicists looked for
mutations in the human genetic code.

An obvious response to the growing threat of nuclear destruction,
whether it should come in a definitive flash or by insidious stages, was to
demand a halt to the arms race and to nuclear testing. A substantial
number of Americans (joined by persons in other countries) did precisely
this. The most visible and articulate of the anti-nuclearist groups was the
National Committee for a Sane Nuclear Policy. SANE, as it was called,
summarized its purpose in a manifesto in the *New York Times* signed by a
distinguished charter membership of physicists, philosophers, writers,
and assorted political activists. "We are facing a danger unlike any danger
that has ever existed," the manifesto asserted. "In our possession and in
the possession of the Russians are more than enough nuclear explosives to
put an end to the life of man on earth." Ominously, man's political development hadn't kept pace with his technology, and the anachronistic competition of nations might terminate competition and nations alike. As a first
step, the nuclear powers should halt weapons testing immediately, and
allow the people of the earth to breath unpoisoned air and cultivate uncon-
taminated soil. A more permanent solution to this unprecedented existential problem, however, required a fundamental rethinking of human priorities. In particular, individuals must develop a loyalty higher than national
loyalties, and must stop conceiving of world affairs in terms of competing
sovereignties. "The sovereignty of the human community comes before all
others."[16]

The SANE manifesto at once identified the essence of the nuclear
problem and indicated its intractability. While at the surface the nuclear
predicament was a case of technology tearing off with human fate, at heart
it represented a failure of politics. Put otherwise, the problem lay less in
the technology than in the technology's configuration of ownership.
Tribes and nations had warred on each other for tens of millennia. In the
process, lots of bystanders were killed and maimed. Now, through the
possession of nuclear weapons, the nations could war more swiftly and
thoroughly than before, and the scope of their extraneous damage was
comparably increased. But, at base, it was the same old story: people
against people, nation against nation. As long as nations enjoyed first
claim on the loyalties of their inhabitants, the story wouldn't change
much—short of a collective and irrevocable finis.

The higher loyalty the SANEists called for might have had a chance of
developing in America in the early postwar years, before the Cold War

consensus congealed and stifled the renascent internationalism. But after ten years of characterizing the communists as hell-bent on world conquest, American leaders showed no interest in power-sharing with Moscow. Eisenhower, for all his concern at the way the arms race was stealing resources needed to improve the lot of struggling humanity, couldn't bring himself to risk losing the American lead. Writing to a friend who had expressed dismay at the lack of progress toward arms control, Eisenhower pinned the blame squarely on the Soviets. "Do you not ignore completely in your discussion," he asked, "the central fact of an almost religious commitment in the Soviet Russian mind to the eventual extinction of our political system and way of life, to be achieved by military means if other methods fail? This country has repeatedly offered a sensible proposal which would lead to complete and total disarmament. Every time such an offer has been made, the Russians have flatly refused cooperation."[17]

Eisenhower subscribed to the two-worlds ideology as thoroughly as any of the original architects of the Cold War. In 1947, the future president summarized his thinking on world affairs in his diary. "The main issue," he wrote, "is dictatorship versus a form of government only by the consent of the governed, observance of a bill of rights versus arbitrary power of a ruler or ruling group. That the issue is with us needs no argument—the existing great exponent of dictatorship has announced its fundamental antagonism to all sorts of capitalism (essential to democracy) and that it will strive to destroy it in the world." Eisenhower continued, "Everywhere the sullen weight of Russia leans against the dike that independent nations have attempted to establish, and boring from within is as flagrantly carried on as is obstructionism in the UN. Russia is definitely out to communize the world—where it cannot gain complete control of territory, as it has in Bulgaria, Poland, Rumania, Yugoslavia, and the Baltic States, it promotes starvation, unrest, anarchy, in the certainty that these are breeding grounds for the growth of their damnable philosophy."[18]

Eisenhower conceded that the Kremlin might disguise its ultimate objective behind rhetoric of reasonableness and coexistence—engaging in what aide C. D. Jackson liked to call "Leninist zigzags"—but he never wavered in his view that the Soviets intended global conquest. In the second volume of his memoirs, published in 1965, he asserted, "Despite difficulties that have intermittently arisen in recent history among free nations, the truly virulent problems in international affairs spring from the persistent, continuing struggle between freedom and Communism." He concluded, "Communists embrace every kind of tactic to gain their fundamental objective: the domination of the earth's peoples." Holding such opinions, Eisenhower wasn't the man to hazard American security on disarmament schemes that allowed the slightest chance for Soviet cheating.[19]

Neither was John Foster Dulles. The secretary of state's views on the Soviet Union, like his views on nuclear strategy, were more sophisticated than he often made them sound. Charles Malik, the Lebanese foreign minister, saw through Dulles. "In the case of Mr. Dulles," Malik remarked later, "he spoke rather categorically, stiffly, in terms of black and white, in terms of good or evil, in terms of God and Satan, and all that sort of thing. That's quite true. But few men, I tell you, had as much flexibility, as much power of nuance, of making nuances, as Mr. Dulles did." Dulles understood how countries, especially in the developing world, might freely—unintimidated, unsubverted, unduped—choose communism over capitalism, simply on the performance of the opposing systems in delivering the goods people needed to survive. He wrote to C. D. Jackson, "I have become personally convinced that it is going to be very difficult to stop Communism in much of the world if we cannot in some way duplicate the intensive Communist effort to raise productivity standards. They themselves are increasing their own productivity at the rate of about 6 percent per annum, which is twice our rate. In many areas of the world such as Southeast Asia, India, Pakistan and South America, there is little if any increase. That is one reason why Communism has such great appeal in areas where the slogans of 'liberty,' 'freedom,' and 'personal dignity' have little appeal."[20]

But Dulles didn't intend to stick his neck out explaining to Republican conservatives and other committed distrusters of the Soviet Union that the divide between East and West was less distinct than they wanted it to be. He concluded his letter to Jackson, who had been pressing the secretary to support increased American aid to India and other Third World countries, by saying that under present political conditions such support was "not practical."

Nor did Dulles intend to campaign for disarmament. The fact that disarmament sat poorly with American conservatives probably would have sufficed to warn Dulles away, but a second consideration confirmed his opposition. Dulles contended that an arms race played to American economic strength. However fast the communists might be raising their productivity rate, their absolute output still trailed America's by a wide margin. If the United States kept up the pressure, an arms race would eventually knock the Kremlin to its knees. "Our economic base, almost equal to that of all the rest of the world together, can support indefinitely the high cost of modern weapons," he wrote. "The Soviet bloc economy cannot indefinitely sustain the effort to match our military output, particularly in terms of high-priced modern weapons. Already there is evidence that the Soviet economy is feeling the strain of their present effort and that their leaders are seeking relief." Rather than grant the relief, Dulles declared, the United States should "continue the pressure."[21]

Eisenhower's and Dulles's disinclination to disarm gained support from some of the country's most eminent scientists. Edward Teller, a prime mover behind the development of America's hydrogen bomb—not to mention a Hungarian expatriate who despised the Soviets for their treatment of his homeland—denounced the disarmament campaign and denigrated the arguments of the anti-testers. As long as the Soviet Union had nuclear weapons, Teller asserted, the United States must have nuclear weapons. And if the United States was going to have nuclear weapons, it needed to test them, in order to ensure the performance of current weapons and to develop new and better ones. As something of a sop to those who feared the effects of fallout, he claimed that further testing would allow progress toward a "clean" bomb that produced terrific concussion and heat but almost no radioactivity. At the same time, though, he extolled the virtues of nearly the opposite of a "clean" bomb: a cobalt warhead that produced almost entirely radiation. But clean or dirty, Teller declared that the United States needed to continue testing in order to guarantee the safety and therefore potency of the American deterrent. "The danger of the test," he stated, "is nothing compared to the catastrophe that may occur if great numbers of these weapons should be used in an unrestricted nuclear war."[22]

Eventually, Eisenhower judged that the pressure for an end to testing required a response, and he agreed to a twelve-month moratorium, later extended. But in other respects, he accomplished nothing significant in the direction of arms control.

In particular, he resisted the sort of anti-nuclear internationalism advocated by the SANEists, preferring a policy of nuclear nationalism. The policy made strategic sense, in a psychological way. Although the "missile gap" Eisenhower's Democratic critics decried existed only in the bombast of Khrushchev and in the imaginations of his gullible listeners, by the late 1950s the Soviets undeniably possessed the ability to inflict appalling damage on the United States. Because no military defense existed against the Soviet threat, American leaders were forced to rely on a psychological defense. The Soviets wouldn't shoot their missiles and destroy the United States because they knew, or believed, or at least worried, that the United States would shoot back and destroy the Soviet Union. Unlike the internationalist solution to the nuclear problem, this version required no assumption of good will or cooperation on the part of the Kremlin. It required only an assumption of Soviet self-interest, which Americans had conceded to the communists in spades since 1917. At the same time, it required the continuing development of the American nuclear arsenal. Deterrence would deter only so long as the United States possessed the capacity to level the Soviet Union even after absorbing a Soviet attack.

The policy of nuclear nationalism had three important consequences. First, it spurred the arms race, since one could never have too much deterrent capacity. American officials could, at best, merely guess the size of the Soviet arsenal. (Reconnaissance by American spy planes allowed reasonably accurate guesses, but until 1960 the U-2 flights were secret from the American people, and the information they produced didn't enter public discourse.) Prudence required estimating high, as the missilegappers did, and building accordingly. Besides, one never knew when the Soviets would discover some diabolical device that would place the American deterrent force at risk. Better, therefore, to put the country's nuclear eggs in more than one basket—ideally giving some to each of the military services. Thus was born America's nuclear triad, which made everyone in the Pentagon happy, generated jobs and profits in the defense industry, and may or may not have enhanced American security.

A second consequence was the closer wedding of the American people to the cause that ostensibly necessitated the weapons race. Americans recognized the utter peril that arming the world on an unprecedented scale was placing them in. They felt the economic burden of maintaining the most powerful and expensive military establishment in human history. Recognizing the peril and feeling the burden, they naturally came to believe that it was all necessary. How else to explain it? In those pre-Vietnam, pre-Watergate days, Americans were inclined to trust what their leaders said on matters relating to national security. Political factions and parties might differ regarding precisely what sorts of weapons to build, and how many, but few persons contested the ultimate need to arm against the communist threat. And by this stage of the Cold War, and of the arms race, each side—American and Soviet—could credibly claim it was simply responding to measures initiated by the other.

The third consequence of the nuclear-nationalist approach was the creation of a new realm of intellectual activity. Since Socrates's time, intellectuals have often been the bane of governments' existence. Intellectuals usually think they know better than governments how governments ought to conduct themselves. Often the intellectuals are right, which perhaps says less about intellectuals than about governments. Regardless, keeping intellectuals from questioning too closely is an essential task of all governments that aspire to remain in power, as all do. In the United States, where hemlock is not an option, government has tried various means to keep intellectuals (and other questioners) from causing trouble. In the 1790s, Congress passed the Alien and Sedition acts to smother those who challenged the Federalists' handling of relations with France and Britain. Congress adopted a gag rule in the 1830s to prevent consideration of the slavery issue. During the era of Grant, government simply overwhelmed naysayers with corruption and triviality beyond comprehension. Congress revived the idea of sedition in 1918, opening the way for the first and second Red Scares of the first and second postwar periods. These inquisitions functioned well in silencing dissent.

But repression is not as effective, over the long term, as co-option. Sooner or later, someone remembers the First Amendment, and the courts summon the courage to demand its enforcement. Clever presidents and administrations pre-empt criticism by giving potential critics a piece of the action. (Lyndon Johnson, who definitely did not succeed in silencing the intellectuals, nonetheless aptly summarized the strategy when he said, regarding one critic, that he'd rather have the man inside the tent pissing out than outside the tent pissing in.) Woodrow Wilson applied the strategy to Walter Lippmann and other liberals who had begun to doubt Wilson's commitment to progressive war aims, by making Lippmann and the others part of the administration's peace-planning "Inquiry." Franklin Roosevelt brought intellectuals into his "brains trust," thereby fostering the impression that the New Deal had more forethought about it than it did. John Kennedy would become the darling of the intellectuals by appointing a few to key positions, several more to not-so-key positions, and generally humoring the rest.

As it applied to national security policy in the Cold War, the co-option gambit produced a new form of intellectual employment: nuclear scholasticism. Because deterrence could never be practically tested without placing America and the world at unacceptable risk, it had to be tested in theory. Because the theories the testers developed couldn't usually be disproved, short of nuclear war, theories begat new theories in profusion. The devising of theories and counter-theories regarding nuclear war became an important growth industry during the 1950s and after. To a certain extent, it attracted hard-science types who had actual experience with the kinds of technology modern weapons entailed. But generally these real scientists didn't need the work, having plenty to do building the weapons. For the most part, therefore, the analysis and consulting industry attracted social scientists and others who specialized in producing words rather than material products. Some of the analysts worked directly for the government, usually in one of the military services, the CIA, or another intelligence agency. Some worked in research institutes that relied on government contracts. Here the prototype was the Rand Corporation, which started life in the late 1940s as a prime intellectual contractor for the United States Air Force. Some worked in universities, often on projects approved and funded by the Defense Department. Many academicians consulted part-time for the government. More than a few entrepreneurial types began in academia or on the staffs of research institutes, and then hung out their own shingles. Start-up costs were minimal, and one's reputation was portable.

The most visible of the nuclear scholastics was Herman Kahn, a Rand alumnus who left to form his own think tank, the Hudson Institute, upriver from Manhattan. While many of the breed contented themselves with narrowly defined and carefully footnoted projects, Kahn thought much bigger. He described his initial five-year plan for the Hudson Institute with typical immodesty: "The first year we would put down on paper everything that a good secretary of defense would want to know; the next year a good secretary of state; the next year a good president; the fourth year a good secretary-general of the United Nations; and the fifth year a good God. And then we'd quit." The project took longer than anticipated. The first three years stretched into several more than that,

leaving Kahn and company stuck working on Washington, and requiring
the secretary-general and God to look elsewhere for advice. All the same,
Kahn made quite a splash. In a series of books beginning with *On Thermonuclear War* (a neo-Clausewitzian primer for the atomic age), continuing
through *Thinking About the Unthinkable* (an expansion of the more provocative points of the earlier volume), and concluding with *On Escalation* (a
how-to guide for nuclear warriors), Kahn brought Americans face-to-face
with the sobering possibilities inherent in the possession of large numbers
of nuclear weapons.[23]

Not everyone took Kahn seriously. At least one reviewer thought the
author intended *On Thermonuclear War* as satire. Another accounted the
book—which included terms like "wargasm," denoting the essentially in-
voluntary climax of escalation—a nuclear version of pornography. Certain
of Kahn's ideas—the "doomsday machine," for instance—found fame in
the black-comedy film *Dr. Strangelove*.

As he no doubt intended, or at any rate expected, Kahn's scenarios
elicited counterscenarios, and the defense-analysis industry continued to
expand, often into areas questionably related to national security. The
activity absorbed a sizable portion of America's intellectual resources, in
the process creating a numerically small, but inordinately articulate group
with a stake in the maintenance of the Cold War status quo. Some of the
more imaginative think-tankers might hope to wangle continued funding
should peace unaccountably break out—Kahn himself went into a less
forbidding form of futurology—but nothing matched national security as
the open sesame of government coffers. Admiral Hyman Rickover explained a simple fact of life to a congressional committee: "The DOD
[Department of Defense] has been able to involve itself in research having
only the remotest relevance to the problems encountered by the armed
services—matters at no previous time nor anywhere else in the world
deemed to lie within the province of the defense function—just because it
has the money. It has more money than any other public agency. It gets
more money because the word 'defense' has in itself an element of urgency.
Whatever is asked in its name somehow acquires the connotation of a life
and death matter for the nation." Rickover added, "There is now virtually
no subject that the DOD might not wish to 'research' on its own, and
virtually no one to gainsay its right to do so. To use an analogy with
nuclear science and technology: As you know, one has to get a 'critical
mass' before a nuclear bomb 'takes off.' What has happened to the DOD, I
think, is that it has attained that 'critical mass' and 'taken off.' " Rickover
remarked that the individuals who evaluated the merit of research proposals were precisely the same types who did the research. "So you have a
sort of interlocking directorate composed of DOD advisers on research
and DOD recipients of research grants—on the one hand those within the
DOD who advocate projects; on the other the university professors, the

universities, and above all the 'think tanks' who get a large share of DOD research and development money."[24]

Open skies

The nuclear-nationalist approach to American security produced further adventures in American covert operations. The most spectacular of these—in terms initially of engineering audacity, and later of diplomatic maladroitness—was the U-2 reconnaissance program. What made the U-2 possible was a series of bright ideas in American aircraft design and photographic technology. What made it seem necessary was American ignorance regarding the rapidity of Soviet nuclear development. The latter was the more crucial. In the absence of hard evidence on Soviet missilery, every group with an interest in defense spending could hire its own analysts and conjure up whatever estimates served its purposes. Eisenhower, who understood the politics of defense better than any president since George Washington, sought solid numbers, to ensure American security and to prevent the Pentagon and its Gaitherist allies from breaking out of the spending restrictions he still tried to impose.

Eisenhower first tried to obtain the evidence by licit means. At the Geneva summit of 1955 he presented his "Open Skies" plan for reciprocal overflights of the Soviet Union and the United States by American and Soviet aircraft. The plan wasn't entirely serious, in that it contained little to make it attractive to the Kremlin. Like the 1946 Baruch plan for international control of atomic weapons, which had offered much less to the Soviet Union than to the United States—granting America effective control over the entire internationalization process, enforceable by means potentially including American nuclear weapons—the Open Skies proposal invited Soviet rejection. As matters stood, the Soviets knew far more about America's military capabilities than Americans knew about Russia's. Allen Dulles once remarked that he would pay "a good many million dollars" for information about the Soviet Union comparable to what anyone, including Soviet agents, could read daily in American newspapers. Open Skies, by leveling the playing field, or airspace, would diminish the Soviets' advantage. Further—and in this respect it paralleled the Marshall Plan—the intrusive inspections it specified would reveal how far the Soviet Union lagged behind the United States. Khrushchev hardly cared to be exposed as the Oz of international relations.[25]

Khrushchev's rejection of the Open Skies proposal didn't especially disappoint Eisenhower. The president gained credit for a reasonable-sounding proposal that if implemented would have lessened tension and reduced the risk of surprise attack. Equally important, he knew that the U-2 would soon begin to open Soviet skies without requiring any concessions whatsoever by the United States.

As has typically been true of American covert operations, the U-2 program was a poorly kept secret. The Soviets learned of it at once, with Soviet radar picking up the planes and tracking them across Russian territory. American allies supplied support facilities and shared the intelligence harvest. The American press, including such influential organs as the *New York Times*, the *Washington Post*, and *Model Airplane News*, had much of the story by 1958. But the big players of the media patriotically held the presses on the program, and no one but hobbyists—probably not even KGB agents—read the *News*.

Again as with most covert operations, the various devices used to keep the program quiet—explanations about wayward weather planes and the like—had the primary purpose of keeping the American public in the dark. What Americans didn't know, and the American government could deny, wouldn't hurt them or politically embarrass the government.

But what they did know, or learned, rather, after the Russians shot down Francis Gary Powers's U-2 in May 1960 caused enough embarrassment to cover retroactively the years of blissful ignorance and deniability. Eisenhower compounded the embarrassment by trying to lie his way out of the predicament. When Khrushchev produced the live Powers, further prevarication became impossible. Eisenhower, now forced into the position that inevitably catches presidents when covert projects go bust, had to choose between appearing ignorant and appearing deceitful. He opted for honest deceit, admitting that he had authorized this violation of international law.

Eisenhower then attempted to shift the blame for the violation onto the Soviets. He reminded the American people that he had proposed and Moscow had rejected the Open Skies plan. Moscow's rejection, he claimed, had required the United States to look to its own devices and its own safety. "No one wants another Pearl Harbor. This means that we must have knowledge of military forces and preparations around the world, especially those capable of massive surprise attacks." The forces of only one country fit this description, of course. And the Soviet Union made the greatest effort to conceal its military preparations. "In the Soviet Union there is a fetish of secrecy and concealment." Indeed, the Kremlin convicted itself by its handling of the present affair. "The emphasis given to a flight of an unarmed nonmilitary plane can only reflect a fetish of secrecy." Facing this fetish, the American government had no choice but to try to crack the Kremlin's stone wall of silence. "Our deterrent must never be placed in jeopardy. The safety of the whole free world demands this."

Eisenhower went on to address the larger topic of espionage in the Cold War. "Ever since the beginning of my administration I have issued directives to gather, in every feasible way, the information required to protect the United States and the free world against surprise attack and to enable them to make effective preparations for defense." Such activities, by their nature, had "a special and secret character" and operated by "their own rules and methods of concealment which seek to mislead and obscure."

The American government and the American people didn't relish operating by such rules. "We prefer and work for a different kind of world, and a different way of obtaining the information essential to confidence and effective deterrents." But that world hadn't yet arrived, and in the meantime the sorts of activities recently revealed remained "a distasteful but vital necessity."[26]

The immediate consequence of the U-2 imbroglio was the wrecking of a summit conference scheduled for Paris just a few days later. Khrushchev went to Paris only to denounce the United States for international banditry and to withdraw an earlier invitation to Eisenhower to visit the Soviet Union. Eisenhower put a brave face on the affair, but was disappointed that the last months of his administration had come to this. George Kistiakowsky, the president's scientific adviser, spoke with Eisenhower after the Paris flop and recorded the substance of the conversation. Eisenhower remarked that, regarding the U-2, his scientists had failed him. Kistiakowsky reminded the president that the scientists had warned repeatedly that sooner or later the Soviets would knock one of the spy planes down. The fault, Kistiakowsky said, lay with the management of the program. Eisenhower's sensitivity on the issue showed immediately. "The President flared up," Kistiakowsky wrote, "evidently thinking I accused him, and used some strong uncomplimentary language. I assured him that my reference was to the bureaucrats that ran the show. Cooling off, the President began to talk with much feeling about how he had concentrated his efforts on ending the Cold War, how he felt that he was making big progress, and how the stupid U-2 mess had ruined all his efforts. He ended very sadly that he saw nothing worthwhile left for him to do now until the end of his presidency."[27]

From Girondists to Montagnards

A tour of the Soviet Union would have made a nice send-off for Eisenhower. It might have allowed an easing of superpower tension and perhaps a slowdown in the arms race as well. It might also have provided that miniscule extra portion of votes the Republicans needed to hold the presidency. But Eisenhower didn't go to Moscow. The Cold War and the arms race continued. And Richard Nixon didn't move into the White House. The retightening of tension during the summer of 1960 provided John Kennedy just the grip the Democratic candidate needed to get to the top of the greasy pole in November.

Kennedy's election and the events that followed revealed an important and superficially paradoxical principle in American politics during the Cold War: If you wanted more of the same in American foreign affairs, you should have voted for the progressive party, while if you preferred change, especially in a liberal direction, you should have voted conservative. Ever since the Republicans had hounded Harry Truman from office

following the communist victory in China and the stalemating of the war
in Korea, Democrats felt the need to prove their toughness against communism. The one Democratic president who initially violated this rule—
Jimmy Carter—changed his mind under relentless pressure from the
right. Republicans, on the other hand, with their anti-communist bona
fides in generally good order, enjoyed greater room for maneuver. Eisenhower could travel to Geneva in 1955, something a Democrat couldn't
have done without rekindling the Yalta debate. Nixon as president would
go to China. Ronald Reagan would achieve the first meaningful reduction
in nuclear weaponry between the United States and the Soviet Union.

As a first step toward proving his toughness, Kennedy attempted to
restore credibility to American defense strategy by junking massive retaliation in favor of "flexible response." The latter incorporated ideas that had
gained currency during the latter half of the 1950s, out of the widespread
disbelief regarding massive retaliation. The essential thrust of the new
strategy was that the United States should possess the capacity to react to
foreign pressure or aggression at a variety of levels, beginning with
garden-variety conventional forces, moving up through mini-nuke artil-
lery shells and short-range rockets, and culminating in city-shattering
intercontinental missiles and long-range bombers. This, of course, was
precisely the defense configuration Eisenhower had rejected as unnecessar-
ily expensive. Needless to say, Kennedy didn't have to battle the military
services to implement his new-old approach. Perhaps he realized that the
battle was hopeless. Eisenhower, the towering military figure on the
American political scene, had had all he could do keeping the Pentagon
and its civilian auxiliaries in line, and by the end even he was despairing.
In his celebrated farewell address, the war hero cautioned his compatriots
against the overweening ambition and oversized influence of the "military-
industrial complex." Kennedy, who had never commanded more than a
PT-boat, could hardly expect to do better than the five-star general.

Yet Kennedy had his own plan for getting more bang for America's buck.
The president turned the Defense Department over to Robert McNamara,
the highly acclaimed CEO of Ford Motor Company. McNamara had scored
a brilliant success with Ford's Falcon, and he seemed an odds-on bet to do
likewise with the Skyhawks, Hornets, Magic Dragons, and other denizens
of the Pentagon's bestiary. To capture the critters, McNamara strode boldly
into the thicket of numbers the bureaucrats customarily considered safe
refuge. Wielding the machete of systems analysis, he slashed and chopped
with relentless vigor. The pruning appeared impressive at first. But as
thickets do, this tangle had a habit of sprouting three shoots where one had
grown before. To complicate matters, by the time McNamara noticed
where he was walking, the ooze of war in Southeast Asia was up to his
ankles. The ooze fertilized the new growth, more than undoing the defense
secretary's work. For all the overhead-trimming, Kennedy's last defense
budget (for fiscal 1964, even before the real escalation in Vietnam began)

was nearly 11 percent higher in real terms than the Eisenhower defense budget of 1960.

If Kennedy revised Eisenhower's New Look policy on nuclear weapons, the Democratic president accepted Eisenhower's inclination to covert warfare. Kennedy arrived in the Oval Office when the CIA's scheme for invading Cuba was nearly complete. During the ten months since Eisenhower had given the agency the go-ahead to form and train a small army of anti-Castro emigres, the operation had gained considerable momentum. In such cases, Eisenhower had reserved the right to veto operations until the last moment. When France had appealed for American help in Indochina in 1954, Eisenhower had allowed the Pentagon to develop plans for an air strike to relieve Dienbienphu, only nixing Operation Vulture at the very end. He might similarly have canceled the invasion of Cuba if the eleventh-hour briefers failed to make their case.

But what Eisenhower, the Republican military expert, could have done, Kennedy, the Democratic novice, couldn't, at least not without great difficulty. As a candidate for the presidency, Kennedy had lashed the Republicans for insufficient vigor in the face of the communist threat. As president, he didn't wish to expose himself to the backbiting leaks a decision against the Cuban operation would certainly entail. The leakage might spring from within his own administration, where the gung-hos could hardly wait for the exile army to hit the Cuban beaches. Or it might come from among the hundreds of Cubans who would have to be demobilized, their hopes of returning to Cuba frustrated. Some leakage was occurring already. A week before Kennedy's inauguration, the *New York Times* ran a front-page article on one of the CIA training camps.

If Kennedy decided against the operation, he would find himself in a position similar to the one Truman had faced after the communist victory in China, with the difference that China was large and far away, making an anti-Mao effort difficult, expensive, and not obviously critical to America's core security interests, while Cuba was small and close, making an anti-Castro effort comparatively easy, inexpensive, and directly related to the defense of the Caribbean and the safety of the Panama Canal. Additionally, by the time China became a big political problem, Truman had already had four years in office and had accomplished much of what he aimed to do. Kennedy, on the other hand, had just entered office. Whatever he would accomplish lay ahead.

Consequently, the president authorized the CIA to send its refugee battalion toward the principal landing site at Playa Giron on the Bahia de Cochinos, or Bay of Pigs. The operation misfired from beginning to end, leaving Castro more of a hero than before to most Cubans and other Third Worlders, and Kennedy red-faced and more sensitive than ever to Castro's presence across the Florida Strait.

Kennedy manfully accepted responsibility for the fiasco—not that he had much choice. He then proceeded to sack CIA director Allen Dulles.

The fiasco didn't cure the Kennedy administration of the desire to over-throw Castro, however. If anything, the humiliation intensified it. Attor-ney General and First Brother Robert Kennedy assumed charge of the anti-Castro program, declaring that solving the Cuban problem was the "top priority" of the administration. "All else is secondary," he said. "No time, money, effort, or manpower is to be spared." Richard Bissell, the CIA deputy director for plans, returned to his office one day in the fall of 1961 and related to a colleague how he had been, as the latter recalled, "chewed out in the Cabinet Room of the White House by both the Presi-dent and the Attorney General" for "sitting on his ass and not doing anything about getting rid of Castro and the Castro regime."[28]

The agency quickly started doing something. It provided weapons, other equipment, and training to Cuban emigres who undertook raids against their home country. It sabotaged cargoes of sugar leaving Cuba, and of foreign goods bound for Cuba. And it resurrected plans for killing Castro. CIA agents met with American mobsters to arrange a hit on Castro. CIA experts explored the possibility of developing an exploding seashell that could be planted on the ocean floor in an area where Castro liked to dive. CIA contacts provided poison to a disaffected major in the Cuban army.

The anti-Castro operations continued until October 1962, when, in the midst of the Cuban missile crisis, the Kennedys decided not to compound their problems by further attempts to destabilize Castro. The operations resumed briefly afterward, but petered out as Washington became increas-ingly mesmerized by another revolution—in Southeast Asia.

During the Kennedy years, America's Vietnam problem grew from a modest annoyance to a howling crisis. Following the creation of the anti-Diem National Liberation Front at the end of 1960 and the onset of a serious insurgency against the South Vietnamese government, American efforts on behalf of the Saigon regime increased dramatically. Much of the effort was visible and acknowledged. Top-ranking, mid-ranking, and low-ranking officials of the Kennedy administration regularly jetted to Saigon from Washington with promises and deliveries of American economic and military aid. As the most obvious manifestation of American support, the Kennedy administration raised American troop strength in the country from several hundred to some 16,000.

Yet much of the war was fought under the cloak of secrecy. The CIA dropped agents into North Vietnam, hoping to disrupt life there and diminish Hanoi's ability to support the rebels in the south. The American navy provided electronic cover for South Vietnamese raids against the North Vietnamese coast. Among the most extensive of the covert Ameri-can operations was the mobilization of the Montagnards, as the French had called the inhabitants of the Vietnamese central highlands. Beginning in 1961, the CIA and American army special forces (the Green Berets) outfitted and trained the Montagnards to resist communist infiltration. The program worked reasonably well, and during the next two years more than 40,000 Montagnards enlisted in village militias. A smaller number joined a mobile strike force.

The program had a hitch, though. The lowland Vietnamese who ran the South Vietnamese government had no more love for the Vietnamese highlanders than lowland Scots have often had for Scottish highlanders. The Montagnards reciprocated the distaste. In creating a Montagnard army, American officials tended to give substance to Montagnard desires for autonomy or independence—which was another reason why Saigon looked decidedly askance at the whole idea. In 1964, the regular South Vietnamese army attempted to assert authority over the Montagnards. This action provoked a rebellion among the Montagnards, and generated a full-blown independence movement. When the South Vietnamese forces and the Montagnards met in open combat, American officers and agents had to decide which party to back. Until the American withdrawal from Vietnam, the question remained unresolved. A paper accord granted the Montagnards greater self-government, but Saigon often pleaded exigencies of war and ignored it.

The arming of the Cuban emigres and the Montagnards marked a new turn in the Cold War, or at least a new twist on an old idea. In each case, the United States trained a secret army of foreigners to do America's bidding against the communists. Presumably, the invaders of Giron and the Montagnards entered the compacts with their eyes open. Yet even so, the result was the creation of a sizable body of foreigners who had a legitimate moral claim on the United States. (To this category one might add the Meo, or Hmong, of Laos, whom American agents were organizing for a Montagnard-like operation against the communist Pathet Lao.) The individuals involved had risked their lives and much else in the fight against communism. For the United States to abandon the fight, if only in their part of the world, would leave them seriously in the lurch. The Cuban exiles were physically safe in Florida and elsewhere, but any rapprochement with Castro would crush the exiles' hopes for regaining what had been lost back home—and would create political problems for the American administration that attempted the maneuver. The Montagnards (and the Hmong) were more at risk, since abandonment of the fight against Southeast Asian communism would lay them open to reprisals. Of the various arguments used over the years to justify continued American participation in the war in Southeast Asia, one that consistently pricked the American conscience was the contention that withdrawal would produce a bloodbath among those who had taken America's side. Americans discovered that allies in the fight against communism could become hostages to America's continuing the fight.

Mano a mano

Perhaps the most remarkable aspect of the Cold War was the success with which the superpowers, despite their avowed hostility toward each other, and their competition in most areas of international affairs and in most parts of the world, managed to avoid direct confrontation. At the beginning of 1946, the Truman administration had felt obliged to indicate its displeasure at the Kremlin's procrastination in withdrawing Soviet troops from Iran. In 1948, Washington let Moscow know it didn't much like having to airlift food, coal, and everything else to the Western zones of Berlin. The Soviets again applied pressure on Berlin in the late 1950s and in 1961. During various of the Middle East wars, Washington and Moscow cautioned each other not to try to take advantage of the regional troubles. Following the Russian invasion of Afghanistan in 1979, Jimmy Carter warned against any further Soviet thrust south. But none of these incidents brought the United States and the Soviet Union especially close to war, and even these were few enough during more than forty years of rivalry.

The one jarring exception to this general rule of avoidance of dangerous superpower-versus-superpower confrontation was the Cuban missile crisis of 1962. The blundering began when the Soviets attempted to steal a march on Washington by slipping medium-range missiles into Cuba. The move, even if completed, would have had little effect on the strategic balance between the superpowers, since the Kremlin would still have lacked the capacity for a one-punch knockout of the United States. Though missiles in Cuba would diminish the reaction time available to American leaders, the difference over missiles launched from the Soviet Union wouldn't be enough for Soviet leaders to count on catching the Americans flat-footed. On the other hand, the missiles *would* have given pause to an American administration plotting another invasion of Cuba. This was what made them appealing to Castro.

But if the Cuban missiles weren't a great threat to America's strategic posture, they *were* a great threat to Kennedy's political standing. To date, Kennedy had distinguished himself in foreign affairs chiefly by making a thorough mess of the Bay of Pigs and by doing nothing concrete to protest the all-too-concrete building of the Berlin Wall. (That there was little he could reasonably have done about the wall didn't change the fact that it damaged the no-nonsense image he desired to project.) At a 1961 summit conference with Khrushchev in Vienna, the Soviet leader had tried to bully the young American president. Although the details of the meeting between the two remained uncommon knowledge, few watching doubted that something of a test of wills had taken place. The midterm progress reports on the Democrats—namely, the 1962 congressional elections—were weeks away, and the Democratic president could scarcely step back from another challenge.

If Kennedy wouldn't step back, neither would his advisers. However bright, determined, and forceful the individuals who make up a president's team are when they arrive in Washington, they soon discover that all their gifts are no match for the fact that only one person in the administration wields real power. Few presidents are secure enough in this power to genuinely appreciate counsel contrary to their inclinations. Even fewer advisers are willing to accept the designated-devil's-advocate status that accompanies consistent objection. With everyone else, presidents like to hear that they are doing the right thing, and advisers quickly learn that influence rubs off on those who tell presidents what they like to hear.

In the case of the Cuban missiles, advisers such as Adlai Stevenson, who recommended a diplomatic solution to the crisis, perhaps involving a swap of admittedly obsolescent American missiles in Turkey for Soviet missiles in Cuba, found themselves labeled appeasers and worse. Although Kennedy rejected the most hawkish proposals—to bomb the missile sites at once and without warning—the president adopted a position hardly more accommodating. He delivered a public ultimatum to Khrushchev to get the missiles out of Cuba, and announced a naval blockade of the island. He reserved the right to take further action in the event the Kremlin failed to comply.

After a few days of bated breath all over the world, Khrushchev gave in. Facing America's clear superiority in nuclear weapons, the Soviet leader decided to press the issue no further. In exchange for a public American pledge not to invade Cuba, and private assurances that the United States would withdraw its missiles from Turkey, Khrushchev ordered the Cuban missile sites abandoned and the weapons withdrawn.

The experience yielded three lessons—one that marginally mitigated the strains of the Cold War, two that reinforced them. The mitigator operated in both the United States and the Soviet Union, and resulted from the fright the brush with nuclear war engendered in most of those concerned. (Only most, not all. Air force general Curtis LeMay, who had been itching for a strike against the Castro gang, found the outcome distinctly disappointing. "We lost," LeMay lamented the day after Khrushchev accepted Kennedy's conditions. "We ought to just go in there today and knock them off.") The crisis added impetus to efforts to alleviate at least the side effects of the weapons race. These efforts led to a 1963 treaty banning nuclear-weapons tests in the atmosphere. The partial test ban had no measurable effect in slowing the arms race, since each side simply took its weapons tests underground. In fact, by depriving the anti-nuclearists of the fallout issue, it reduced the pressure for disarmament. But it did demonstrate the possibility of superpower agreements regarding nuclear weapons, and it cleaned up the air and water breathed and drunk by the billions of bystanders.[29]

The second lesson of the missile affair applied to the Soviets. Having been humiliated by the Americans, the Kremlin determined not to be caught holding the short end of the nuclear stick again. The Soviet counterpart to the American military-industrial complex used the Cuban affair much as the Pentagon and its private-sector allies had used the Sputnik flap to refocus American attention on military and scientific preparedness.

Russia's weapon builders demanded and received a large portion of Soviet national resources, with the consequence that by the early 1970s they had achieved effective parity with the United States in planet-demolishing power. Had it occurred to Richard Nixon to try to intimidate Leonid Brezhnev as Kennedy had intimidated Khrushchev, the situation, dicey enough in 1962, would have been far dicier. Not surprisingly, the Soviet weapons buildup prompted demands in the United States for additions to the American arsenal. These demands proved effective, spurring the Soviets to build all the more. The wheel continued to turn.

The third lesson, in reality ironic but nonetheless critical in American thinking, was that toughness pays. Though Kennedy publicly took the position that the Kremlin's sneaking missiles into Cuba constituted an intolerable provocation, privately he wondered whether it was all that big a deal. "What difference does it make?" he mused at a top-level meeting early in the crisis. "They've got enough to blow us up now anyway." And despite his public resistance to a Cuban-missiles-for-Turkish-missiles swap, he conceded that Khrushchev had a strong case for demanding such a deal. "He's got us in a pretty good spot here," the president said as the crisis peaked, "because most people will regard this as a not unreasonable proposal." At another point in the same meeting, the president remarked, "To any man at the United Nations or any other rational man, it will look like a very fair trade."[30]

But Kennedy didn't want to make the trade, not openly at any rate, for fear of the political backlash. "After all," he acknowledged, "this is a political struggle as much as military." The president's national security adviser, McGeorge Bundy, explained that pulling the missiles out of Turkey, under the current duress, would demoralize America's allies. "If we accept the notion of a trade at this stage," Bundy said, "our position will come apart very fast." Although Bundy didn't agree with the charges the Turks and others might level at the United States—to the effect that Washington was selling allies down the river in order to protect itself—he couldn't deny the significance of such criticism. "It's irrational, and it's crazy, but it's a *terribly* powerful fact."[31]

For all the private ambivalence and uncertainty, and despite Kennedy's willingness, which Dean Rusk revealed many years later, to make an open swap of the American missiles in Turkey for the Russian missiles in Cuba if matters came to that, Kennedy and his advisers presented a public image of steely resolve. The president possessed the grace and good sense not to embarrass Khrushchev any more than necessary (although the experience certainly didn't improve the Soviet leader's standing among his Kremlin rivals, who deposed him in 1964). Nor was Kennedy the type to flaunt his political machismo the way John Foster Dulles had liked to do in professing his and Eisenhower's willingness to press to the brink of nuclear war. Kennedy preferred understated cool to exaggerated bluster. But the president didn't lack advocates who spread the tale of how he had calmly faced down the Russians. The immediate impact was unmistakable. Kennedy's public-approval rating jumped 12 points between polls taken just before and just after the October crisis. More important for the long term was the myth-making that commenced shortly after Kennedy's assassination a year later. Adviser-admirer Theodore Sorensen had Kennedy placing the odds of war with Russia at "between one out of three and even," yet all the same the president didn't flinch from his duty. Administration historian Arthur Schlesinger exulted over the president's "combination of toughness and restraint, of will, nerve, and wisdom, so brilliantly controlled, so matchlessly calibrated."[32]

For the American leaders who followed him, Kennedy's performance during the Cuban missile crisis—as publicly understood—became the standard against which they had to expect to be measured in any similar pinch. The message was plain: be tough, don't give in, even if the stakes involve the fate of the entire human species. Fortunately for the species, the superpowers steered clear of such harrowing confrontations in the years that followed. But the lesson carried over to smaller faceoffs in Southeast Asia, in the Middle East, in Central America, and in Africa, and the lesson was that if the United States stood firm and tall, the communists would back down.

CHAPTER FOUR

The Wages of Hubris
1962–1968

What Harry didn't know

The year of the Cuban missile crisis saw the fifteenth anniversary of the Truman Doctrine. Truman, a practicing politician whose training had honed his grasp of the here and now at the expense of his appreciation of the there and then, had little idea what he was getting America into with his open-ended pledge of support for anti-communist causes. The nature of the commitment grew clearer during the next decade, as Washington concluded mutual-security pacts with countries all around the globe. But with the major exception of China, and the minor exceptions of northern Vietnam and Cuba, until the early 1960s American leaders managed—by armed force, as in Korea; by subversion, as in Iran and Guatemala; by diplomacy, as in Austria, neutralized by agreement among the great powers in 1955; by moral and financial support, as in the Philippines, where American aid helped the Manila government defeat the Huk insurgency; and by luck, as in Yugoslavia, where Tito had carved out a neutralist niche for his country—to hold or slightly push back the line against communist expansion. On balance, the containment policy had to be judged a success.

Yet success never bred much in the way of American self-confidence. Although Americans during the early 1960s showed fewer signs of collective neurosis regarding communism than they had a decade before, they still generally subscribed to the view that a communist victory anywhere would endanger peace everywhere. And despite the construction of anti-communist barriers around the globe, Americans still felt threatened and beleaguered.

Two factors accounted for this sense of beleaguerment. The first was geometrical—topological, to be precise. It is a characteristic of spherical objects like the earth that a circle drawn on the surface has neither an inside nor an outside: both regions are essentially equivalent. (Large circles such as the equator best demonstrate this property.) Put otherwise, each region surrounds the other. Because of this, it was logically valid for Americans to feel surrounded by the communists even as they were surrounding the communists. One can only guess at the psychological dynamics of a Cold War fought on a flat earth.

The second factor was political. The twenty years after 1945 witnessed an accelerating trend of decolonization in Asia and Africa. The European colonialists handed over power slowly in the 1940s, faster in the 1950s, and in a mad rush in the early 1960s. The result was the creation of a large area of instability in which indigenous groups often thought unkindly of the Western powers. Soviet leaders, sometimes anticipating independence and sometimes following in its wake, attempted to take advantage of the instability and unkindliness by backing leftist movements under the rubric of "national liberation." Communist-led, communist-inspired, or communist-supported insurgencies developed in a variety of locales: against the British in Malaya, against the Dutch in Indonesia, against the Belgians in the Congo, against the Portuguese in Angola, and against the French in Indochina, to name five. (Insurgencies like Castro's in Cuba against the American-supported Batista government could have been placed in a parallel category.) Although by the early 1960s the insurgents had relatively little to show for their efforts, their apparent ubiquity and their undeniable energy rendered plausible Moscow's claims of a dawning age of Third World socialism.

More than a few American officials and opinionists argued for a policy of adapting to, rather than opposing, revolutions. None put the case better than John Kennedy in a 1957 speech condemning the Eisenhower administration's decision to side with France in that country's struggle against Algerian nationalists. "The most powerful force in the world today is neither communism nor capitalism, neither the H-bomb nor the guided missile," the Massachusetts senator proclaimed. "It is man's eternal desire to be free and independent." The great enemy of this desire, Kennedy said, was imperialism, which came in two forms: Soviet imperialism, and—"whether we like it or not, and though they are not to be equated"— Western imperialism. The Eisenhower administration, from its fear of the former, had fastened America to the latter. Heeding warnings from Paris that a negotiated settlement with the Algerian rebels would ruin the French republic, the president and his advisers had backed the French government. This policy was doubly misguided, Kennedy asserted, as the recent history of Southeast Asia showed. "Did we not learn in Indochina, where we delayed action as the result of similar warnings, that we might have served both the French and our own causes infinitely better, had we taken a more firm stand much earlier than we did? Did that tragic episode not teach us that, whether France likes it or not, admits it or not, or has our support or not, their overseas territories are sooner or later, one by one, inevitably going to break free and look with suspicion on the Western nations who impeded their steps to independence?"

Kennedy expanded briefly on the Indochina analogy. "The United States and other Western allies poured money and material into Indochina in a hopeless attempt to save for the French a land that did not want to be saved, in a war in which the enemy was both everywhere and nowhere." He continued, "We accepted for years the predictions that victory was just around the corner, the promises that Indochina would soon be set free, the arguments that this was a question for the French alone."

Returning to Algeria, Kennedy lamented the corrosive effects of the war there on France. The "exhausting conflict" had drained the manpower, the resources, and the spirit of the French nation. It had caused the French government to postpone domestic reforms and to attempt to stifle dissent, provoking the people to turn their anger and frustration against the government. The United States, Kennedy concluded, must distance itself from such shortsighted self-destruction. To this end, he offered a resolution recognizing the "legitimate nationalist aspirations" of the Algerian people, and calling on the Eisenhower administration to work for a settlement guaranteeing the fulfillment of such aspirations.[1]

These were brave words, well suited to an aspirant to leadership of the party in opposition. Yet once in the White House, Kennedy succumbed to the peculiarly Democratic disease of creeping communophobia. Interpreted charitably, his shift to a harder line against Third World radicalism reflected the sobering influence of power and responsibility. Nationalism had its place, but American security came first. The Soviet Union could destroy the United States, and American leaders must oppose anything that augmented Soviet strength, including the installation of communist or otherwise pro-Soviet regimes in Asian and African countries. Interpreted skeptically, Kennedy's rightward shift indicated his understanding that no American president had ever suffered rebuke at the polls for insufficient support of Third World nationalism. On the other hand, every president since 1945—and especially the Democrats—had to worry about charges of softness on communism.

Kennedy was able to finesse the issue of nationalism versus communism in Laos. At the time he entered office, the insurgency of the Pathet Lao seemed the most pressing problem confronting the United States in Southeast Asia. Eisenhower, on his way out the White House door, warned Kennedy that the fall of Laos to communism would open the rest of the region to communist conquest. Unhelpfully, though, Eisenhower also opposed the other two obvious options: American intervention and a negotiated settlement. Kennedy, unlike Eisenhower, had to choose, and he chose negotiation. A 1962 accord fashioned in Geneva created a coalition government that neutralized the country and lasted just long enough for Vietnam to cause most Americans to forget Laos.

Vietnam presented Kennedy and the United States with an unavoidable parting of the ways between nationalism and communism. If hard cases make bad law, they make even worse foreign policy. On nationalist grounds, the American government should have supported Ho Chi Minh, who had been fighting for Vietnamese independence from foreign rule since before Kennedy was born. Indeed, Washington had assisted Ho during World War II. But Ho's communism went back almost as far as his nationalism, and on this ground the United States should have opposed him. It did. Kennedy continued the Eisenhower policy of economic and military aid to the Diem regime in Saigon, and when Diem nonetheless looked likely to fall, Kennedy pushed more chips onto the table, greatly increasing the number of American soldiers in the country. When America's money, weapons, and troops still failed to stem the rebellion against Diem's government, Washington discreetly encouraged disaffected officers of the South Vietnamese military to overthrow their president. Perhaps the accompanying assassination of Diem came as a surprise, perhaps not.

Kennedy's assassination three weeks later came as more than a surprise. Kennedy's death shocked America and much of the world. Lyndon Johnson detected divine retribution in the close coincidence of murders. But life went on for Kennedy's heirs, and Johnson immediately found himself having to decide what Vietnam meant to him and to the United States. In the pattern of political leaders, he sometimes confused the two meanings.

To the United States, Vietnam had never meant much intrinsically. Its natural resources paled next to those of neighboring Indonesia. The Philippines afforded better military-basing facilities. Taiwan was more secure against communist attack. But despite Vietnam's deficiencies, three American administrations had made Vietnam a test of America's anti-communist credibility. Truman, shaken by events in China and Korea, had thrown American support to the French colonialists. Eisenhower had compounded the commitment to Vietnamese non-communism by encouraging the creation of a separate government in Saigon, in violation of the 1954 Geneva accords; by treating as permanent the temporary division the accords had specified at the 17th parallel; and by sponsoring SEATO. Kennedy deepened the commitment further by sending thousands of American troops.

For Johnson to have reneged on the American commitment would have been nearly inconceivable. In the first place, Johnson accepted the two-worlds ideology on which the commitment rested. When he spoke of Vietnam being the forward defense of California, he meant what he said. Second, as the inheritor of the policies of the martyred Kennedy, Johnson couldn't fail to pursue Kennedy's policies—certainly not before getting elected on his own in 1964. Third, Johnson was a Democrat, and as a Democrat he understood the special obligation of Democrats to resist communism. Finally, Johnson had big plans for reforming America domestically. He had seen how the communist victory in China and the uproar that followed had killed Truman's Fair Deal in the cradle. Johnson didn't yet have a name for his Great Society, but he had the vision it would embody. He knew that coaxing Congress to embrace the vision would require all the force and guile he could muster. The last thing he needed was for Vietnam to become a major issue. Nothing would make it an issue faster than a communist takeover of the south.

Nothing, that is, except perhaps an all-out American war effort. A policy of declaring war in Vietnam, mobilizing the reserves, and putting America on an emergency footing was nearly as unacceptable to Johnson as a communist victory. Johnson wasn't old enough to remember how the American entry into World War I had snuffed out the progressive spirit of the first decade and a half of the century, but he had read about it. He *did* recall how the second round of the modern thirty years' war had transformed Franklin Roosevelt from Dr. New Deal to Dr. Win the War. No American president has ever known Congress better than Johnson did, and Johnson knew that Congress could handle at most one major task at a time. He could get the legislature to approve his reform program, or he could get it to concentrate on fighting a war in Southeast Asia. But he couldn't get it to do both.

Johnson opted for domestic reform, in the reasonable—though ultimately mistaken—belief that a superpower such as the United States ought to be able to take care of problems in small countries like Vietnam using only a minor portion of its resources. Johnson would have been right had he and his advisers not overestimated the potential stability of the government of South Vietnam, and had Washington not underestimated the willingness of the North Vietnamese and their comrades in the south to suffer for what they believed in.

The overestimation of South Vietnam revealed a critical weakness of American policy in the Cold War. Despite America's enormous power relative to the rest of humanity in the late 1940s and 1950s, it had never been within the ability of the United States to garrison the entire non-communist world against communist expansion. From the first, American leaders had relied for assistance on other countries. In certain cases, the reliance was well founded. The countries that joined the United States in NATO were, for the most part, as firmly attached to democracy as the United States, and had as strong a desire to defend themselves against communist attack. American assistance, in such forms as Marshall Plan aid and American troops in Germany, supplied the mortar that held the European anti-communist wall together, but the NATO allies supplied the bricks. Beyond the North Atlantic core of the American alliance system, however, the United States often had to supply the bricks as well. Across much of Asia and Latin America, from Iran, Pakistan, and South Korea to Nicaragua and Guatemala, American allies and clients cared almost nothing for democracy, attaching themselves to the United States not out of adherence to American values but out of a desire for American largesse. As long as Washington kept the money flowing, the Pahlavis, Rhees, and Batistas said what was expected of members of the Free World.

As long as they provided bases and listening posts, Washington expected little more than talk.

The lowering of the qualifications for membership in the Free World had at least three ill consequences for the United States. First, it converted the American alliance system from one based on conviction to one based on convenience. To a certain degree, of course, all alliances are alliances of convenience, as America's own experience during World War II demonstrated. Yet alliances in which the allies share certain basic values beyond opposition to a common foe tend to hold together longer than those in which they don't. Once Pakistan, for instance, saw that its American alliance didn't prevent Washington from shipping military aid to India, it had few compunctions about joining hands with China, clearly against India and implicitly against the United States.

Second, by aligning itself with unrepresentative and in many cases extremely unpopular regimes, the United States aligned itself, in effect, against the populations of the countries the regimes controlled. This fact nearly ensured that any uprising against the regime in power would be an attack against the United States as well. The communists who seized control of China in 1949 took their Marxist-Leninist ideology sufficiently seriously that profitable relations with the United States would have been difficult in any case, but previous and continuing American support for Chiang made such relations impossible. Castro in Cuba initially—that is, before his conversion to communism—had no compelling ideological reason for his anti-Americanism, but Washington's backing for Batista provided him ample political and emotional cause.

Third, by allying with repressive regimes, the American government undercut the popular moral base on which America's containment policy rested. Whatever appeals to self-interest American administrations since Truman had made in support of containment, the policy wouldn't have stuck so thoroughly in the American mind had a majority of Americans not considered it the morally correct thing to do. Supporters of "realism" in foreign affairs, from Alexander Hamilton and John Quincy Adams to George Kennan and Hans Morgenthau, have often groaned at the tendency of Americans to view international affairs as an arena for deciding matters of morality. Prudent or foolish, the tendency seems inseparable from American foreign policy—not least because even the most hard-nosedly pragmatic policy-makers have exploited the tendency when it has suited their purposes. Every administration from Truman's to Johnson's, and on to Reagan's and George Bush's, cast the contest with the Soviet Union as a moral conflict. To be sure, Americans could swallow their scruples momentarily when confronted by an emergency, as when they had accepted the lesser-evil alliance with Stalin against Hitler. But if the Cold War was an emergency, it was an inordinately long-lasting one. And as the American government forged alliances with regimes that canceled or rigged elections, and jailed or murdered dissenters, Americans questioned—occasionally at first, but more frequently later—the wisdom of such alliances.

These three ill consequences came together in American policy toward Vietnam. By the time the strain on Saigon had reached the point of necessitating outside intervention, SEATO, the alliance established to safeguard South Vietnam, was scarcely more than a dead letter. Pakistan hadn't quite defected to the Chinese side, but its increasingly close ties to Beijing made it unwilling to participate in an effort to save South Vietnam from communism. The Philippines provided grudging support in the form of a token engineer battalion, but Washington had to pay for this, and for plenty more. In the area that would have done the United States some genuine good—letting American B-52s fly combat missions from bases in the Philippines rather than from more distant Guam—Manila said no. Thailand allowed combat missions to take off from Thai soil, but contributed no troops. With the Asian members of SEATO showing so little concern for South Vietnam's future, the non-Asian members felt scant pressure to join the United States. Australia dispatched troops, but Britain, France, and New Zealand steered clear. Eisenhower in 1954 had declared that he wouldn't send American forces into battle in Vietnam without the company of forces from other nations. At the time, it had seemed a wise precaution. In the 1960s, Kennedy and Johnson sent American troops into battle, essentially without that company. The isolation only grew worse as the conflict escalated, leaving the United States a neo-imperialist pariah in the view of many—probably most—countries.

The second ill consequence of the lowered standards for alliance with the United States—implicit American opposition to the needs and desires of the masses of people in the developing world—contributed, in the case of Vietnam, to the first ill effect, and produced problems of its own besides. The repressiveness of the Diem regime triggered widespread protests that scared off potential allies in America's Vietnam venture. At the same time, it reinforced the perception that the United States was simply the latest in the line of foreign oppressors of the Vietnamese people. The popular anti-Diem movement predictably became anti-American as well. Diem's ouster helped little, since the successor regimes were hardly less repressive and were widely seen as creatures of Washington. American analysts recognized that not all opponents of the Saigon governments were communists, but by supporting these governments the United States gave the non-communist dissidents almost no alternative to joining forces with the communists.

The third ill effect—the sapping of American certainty that what the United States was doing was right—gradually eroded American determination to defend South Vietnam. Relatively few Americans bought the extreme antiwar argument that the United States was engaged in genocide and comparable crimes against humanity in Vietnam—although the number of those who did buy it increased with time. Yet the growing understanding of the undemocratic character of the government of South Vietnam eroded the moral assurance Americans required to stick to their task. Many Americans realized that not all the insurgents could be in Hanoi's pay, and the Buddhist monks who immolated themselves to protest the government's excesses provided incontrovertible evidence of profound dissatisfaction among the people of South Vietnam. Had the war lasted only a few years, the questioning of the justice of America's cause in Vietnam wouldn't have made much difference. But as the war dragged on, and as the sacrifices the war demanded of the American people mounted, more and more Americans asked whether the United States had any business being in Vietnam. Eventually a working majority said no.

One, two, three,
What are we fightin' for . . . ?

Vietnam lay at the end of the line of America's Cold War reasoning, the conclusion of a chronological and syllogistic chain that stretched from the origins of the Cold War to the 1960s. When the conclusion proved wrong, or at least unachievable, many Americans traced back along the chain in an effort to determine where the error had arisen. The backtracking produced the most serious intellectual assault on American foreign policy in a generation—an assault, moreover, that formed an essential part of the most sweeping challenge to fundamental American values since the decade of the Great Depression. But where the old left of the 1930s had fired its salvos against American orthodoxy across the rubble of a collapsed economy, the New Left of the 1960s stormed the barricades amid a domestic prosperity unprecedented in American history. In large part, this difference accounted for the larger emphasis on foreign policy in the New Left critique. Domestic troubles still existed in the 1960s, but they were not as central to the lives of the bulk of the American population as the difficulties of the Great Depression had been. Yet the emphasis on foreign affairs also demonstrated the degree to which the Cold War suffused American life after 1945, making issues of international affairs essential to domestic political debates. During the 1930s, an isolationist wall had separated American domestic affairs from foreign affairs—not completely, but sufficiently that Americans could make a fair try at ignoring what happened overseas. During the 1960s, no such wall existed, although the Johnson administration sought to erect one. The administration's failure contributed significantly to Johnson's political demise. When the New Leftists spoke of "bringing the war home" they were referring specifically to the Vietnam War. The larger Cold War had been at home since the 1940s.

The radical attacks on the Cold War and on the mindset and institutions it generated ran a gamut from the disillusioned liberalism of the soft left to the never-illusioned Marxism of the hard left. Historian William Appleman Williams set the tone for the former with his *Tragedy of American*

Diplomacy, first published in 1959 but extensively revised and enlarged in 1962. To Williams, American foreign relations since the beginning of the twentieth century had been based on a crucial contradiction between the genuine idealism of the American people and the self-interested demands of American capitalism. Americans mentally linked democracy and capitalism as necessary concomitants of each other. Because the two had developed together in the United States, Americans tended to assume they would naturally go together in other countries. American "open door" diplomacy, originally enunciated toward China but extended to American relations with most of the rest of the world, embodied a desire to export these best features, as Americans deemed them, of the political and economic realms.

In practice, however, democracy proved less exportable than capitalism. Where the two came into conflict, Williams said, as they often did in the developing countries, democracy usually lost out. American officials pursued marketing and investment opportunities with far more zeal and forcefulness than they promoted democracy. In Williams's judgment, this was a tremendous tragedy, not least because because it was largely unnecessary. If Americans would only reflect on the consequences of subordinating democracy to capitalism—consequences that had grown particularly pernicious since the onset of the Cold War—they would realize the advantages of a true open-door policy, one that would allow simultaneously international exchange on an equal basis and genuine self-determination. "Isn't it time to stop defining trade as the control of markets for our surplus products and control of raw materials for our factories?" asked Williams. "Isn't it time to stop depending so narrowly—in our thinking as well as in our practice—upon an informal empire for our well-being and welfare?" Williams acknowledged that a policy premised on the best in American values, rather than the worst, entailed certain risks. "To transcend tragedy requires the nerve to fail." But he thought such a policy wouldn't fail. "A positive effort to transcend the Cold War would very probably carry the United States and the world on into an era of peace and creative human endeavor."[2]

Sociologist C. Wright Mills detected less room than Williams for the redemption of American foreign relations. In *The Power Elite*—written, like Williams's *Tragedy*, in the late 1950s but also a seminal text for the revisionist movement of the 1960s—Mills described American life in the Cold War era as dominated by a small group of political, military, and scientific leaders. The military and its accomplices were particularly powerful in matters of foreign policy. "No area of decision has been more influenced by the warlords and by their military metaphysics than that of foreign policy and international relations," Mills wrote. "In these zones, the military ascendancy has coincided with other forces that have been making for the downfall of civilian diplomacy as an art, and of the civilian diplomatic service as an organized group of competent people." The most significant of these "other forces" were the political compulsions that followed from the definition of the contest with the communists as chiefly a military struggle—which definition was at once a cause and a consequence of the military's rise. The prospects for undoing the Pentagonization of American international affairs were not good. "With the elite's acceptance of military definitions of world reality, the professional diplomat, as we have known him or as we might imagine him, has simply lost any voice in the higher circles." So long as the Cold War lasted—and, given the needs, desires, and resources of the military and the military's civilian allies, it appeared likely to last for a long while—the elite likely would continue to exercise a death grip on American foreign policy.[3]

Political economists Paul Baran and Paul Sweezy pushed the radical critique still further left. Where Williams judged the excesses of American foreign policy a tragic mistake, and Mills interpreted them as the manifestations of the desires of a narrow elite, Baran and Sweezy saw America's international errors as the necessary outgrowths of an entire oligarchic-capitalist-imperialist system. Viewing contemporary affairs from an explicitly Marxist perspective, Baran and Sweezy in *Monopoly Capital* asserted that a single motive had dominated American foreign policy—and indeed the foreign policies of the other capitalist nations—since the Bolshevik revolution of 1917. "The central purpose has always been the same: to prevent the expansion of socialism, to compress it into as small an area as possible, and ultimately to wipe it off the face of the earth." After a brief fling with intervention in the Russian civil war, American leaders had tried to strangle the new Soviet state by economic and diplomatic sanctions. The sanctions failed, and after World War II the United States switched to armed containment and counter-revolution. The new approach served both the purposes of the monopolists of the military-industrial sector and the needs—economic and psychological—of the capitalist system as a whole. Economically, the policy produced profits that stabilized the American system, fattening capitalists and co-opting workers. (To the extent that workers realized they were receiving less than their fair share of the profits of their labor, they generally considered employment, even if exploitative, preferable to the widespread unemployment they had suffered during the 1930s.) Psychologically as well, the militarization of foreign policy during the Cold War reinforced the position of the oligarchy. Baran and Sweezy contended that though government funds spent on schools or income support might yield levels of economic activity comparable to those generated by military programs, the latter had special benefits for the oligarchy. The oligarchy knew what it was doing. "Whereas massive government spending for education and welfare tends to undermine its privileged position, the opposite is true of military spending. The reason is that militarization fosters all the reactionary and irrational forces in society, and inhibits or kills everything progressive and humane. Blind respect is engendered for authority; attitudes of docility and conformity

are taught and enforced; dissent is treated as unpatriotic or even treason-
able. In such an atmosphere, the oligarchy feels that its moral authority
and material position are secure."[4]

The work of Williams, Mills, and Baran and Sweezy both inspired and
mirrored a far broader attack on the foundations of American Cold War
policy. Historians examining the origins of the postwar conflict with the
Soviet Union contradicted conventional wisdom on the subject, placing
primary blame on the United States and describing Moscow's actions, if
culpable at all, as understandly defensive and designed to ward off American expansionism. Social scientists investigating the causes of poverty in
Asia, Africa, and Latin America described a debilitating dependency of
the developing nations on the industrial capitalist states, a dependency
promoted by the reactionary governments of North America and Europe
and implemented by soulless multinational corporations.

As America is a chronically political society, it was inevitable that the
attack should proceed from the intellectual sphere to the political. Students of radical professors carried what they learned in the classroom out
onto the campuses, from there into surrounding communities, and thence
into the national political arena. Some worked within established parties,
especially the Democratic party, stuffing envelopes and walking precincts
for Eugene McCarthy, Robert Kennedy, and George McGovern. Others
founded new organizations, such as the Students for a Democratic Society, that deemed the major parties beyond salvation. The most radical
hoped and worked for a revolution that would sweep away the last vestiges
of the Cold Warfare state.

The attack took a cultural twist when the disaffected began creating, or
attempting to create, an alternative to the suburban mentality that seemed
to epitomize all that was wrong with bourgeois America. The counterculture that cropped up in Haight-Ashbury in San Francisco and on com-
munes in Tennessee, and that came together most picturesquely in the
mud of Woodstock, represented, among other things, a thorough rejection
of the values, attitudes, and policies that were bringing the United States
to grief in Vietnam.

If the Cold War could be said to have a life, the challenge mounted by
the New Left constituted a midlife crisis. At no other time between the
late 1940s and the late 1980s was America so close to throwing over the
entire institutional and psychological structure on which America's Cold
War policies rested. Though the faculty-club Marxists, the Weathermen,
and the hippies certainly didn't represent mainstream America, the ideas
and emotions they expressed rippled outward into the wider population.
The ripples eroded the mainstream's banks, shifting the channel to the
left. In doing so, they ate away at the Cold War foundations of America's
approach to the world.

Yet even at the most critical moment of the New Left attack, the American Cold War framework was not particularly close to dissolution. For all

the national soul-searching Vietnam provoked, the majority of Americans found it easier to interpret the unsatisfactory situation in Southeast Asia as the result of a misapplication of essentially sound premises than to view it as the consequence of faulty premises. Relatively few came away from the Vietnam experience thinking that the United States should give up opposing communism. Still fewer accepted the revisionsts' claim that the United States never should have opposed it. Vietnam—corrupt, distant, and, come to think of it, not really all that important to American interests—seemed simply the wrong place for the United States to make a stand. Perhaps America couldn't defend the entire world against communist encroachment. But this needn't imply that America should defend none of the world.

The counterattack commences

To a significant degree, the very vehemence of the New Left assault on the premises of American foreign policy played into the hands of persons who desired to vindicate those premises. By overstating their case, by charging American leaders with responsibility for nearly everything that had gone wrong in the world since 1945, the radicals offered easy targets for individuals and groups who wanted gun them down. The cultural side of the assault on the status quo offered opponents of the New Left even easier targets. While the intellectual critique of American Cold War policies had nothing to do with the drug scene or free love, it was tempting for defenders of the status quo to act as though the separate phenomena were closely tied. An overheated Lyndon Johnson sometimes claimed that confronting communist aggression was like confronting the neighborhood bully. If you didn't stop him at the front gate, Johnson explained, before you knew it he'd be on your porch, and right after that he'd be raping your wife in your own bed. Conservatives didn't usually claim that if the radical professors won the intellectual argument in the classrooms the longhairs would soon have your daughter smoking dope in bed in a commune, but they didn't have to. Traditionalists felt that their whole constellation of values was under attack, which it was. In such circumstances, they didn't always take time to distinguish among the various forms of the attack.

In Hegelian fashion, the leftist thesis produced a rightist antithesis. Actually, the first stirrings of the new conservatism antedated the uprisings of the 1960s, and in this respect the New Right wasn't a response to the New Left. Yet in another respect it was, for though the originators of the new conservatism had been decrying the condition and direction of American life, plotting to take over the Republican party, writing for each other's journals, and reviewing each other's books since the middle 1950s, their beliefs gained a grip on the imagination of a substantial portion of the American public only when American society began to fall apart under the combined strains of Vietnam, urban riots, student demonstrations, the counterculture, and the other forms of the uproar of the 1960s.

If one had to choose a single date to mark the birth of the new conservatism, November 19, 1955, would serve better than most. This date graced the cover of the inaugural issue of the *National Review*, a journal of opinion inspired, edited, published, and owned—all the voting stock, that is—by William Buckley. Buckley proceeded to declare war on the Liberalism (his capitalization) that was running America to ruin (his interpretation). In its place, he proposed what he considered a genuine conservatism based on nearly libertarian free-market principles at home and unrelenting anticommunism abroad. The new journal created a small stir when it first appeared, and promised to create a bigger fuss in the future, although through the rest of the 1950s it served principally as a receptable for excess Buckley monies.

Attracting greater notice for the time being, if in somewhat less sophisticated circles, were the activities of Robert Welch, a graduate of the University of North Carolina and Harvard Law School, and more recently an executive in his family's candy business. Welch deemed the decline of Joseph McCarthy a signal setback for American freedom, and in 1957 he left the world of sweets for that of anti-communist political agitation. He developed one of the more encompassing conspiracy theories of the twentieth century, delineating how the communists had weaseled their way to the absolute pinnacle of American politics. Dwight Eisenhower, Welch declared, had for many years effectively been an agent of the Moscow monolith. In an early, privately circulated version of his manuscript, Welch stated that Eisenhower was conscious of what he was doing. In the public edition, the author backed off slightly, conceding that Eisenhower might be an amoral opportunist, "entirely without principles and hungry for glory," or a simpleton "too dumb to understand" the significance of his actions on behalf of the reds. But the result was the same. "We have a Communist, or a politician who serves their purposes every bit as well, sitting right in the chair of the President of the United States."[5]

To alert the American people to the danger in their midst, and otherwise to disseminate his views, Welch founded the John Birch Society. The Birchers opened bookstores, ran booths at county and state fairs, rented advertising space, established local chapters, and enrolled members by the scores of thousands. Sometimes they worked in political campaigns, although most Republicans and nearly all Democrats were too liberal for their tastes. They mounted efforts to remove the worst of the liberals from office. Their billboard campaign to impeach Chief Justice Earl Warren helped relieve the tedium of long stretches of rural highways.

The Birchers were an embarrassment to some new-conservative types, but by no means to all of them. William Buckley devoted an article in the *National Review* to defending the Birchers against what Buckley judged the nervous assaults of the left. "Why is there so much interest in the John

Birch Society?" asked Buckley. Because conservatism was gaining adherents, he answered, and the leftists were worried. "The Liberals, and to the extent their programs coincide, the Communists, feel threatened by the revived opposition." Buckley accounted the objectives of the Birch society laudable. "Its principal aims are to arrest the Communist conspiracy, and resist the growth of government." Buckley conceded that he differed with Robert Welch on some issues—regarding Eisenhower's role in the conspiracy, for example—but he thought the most mistaken of the Birchers' projects less pernicious than those promoted by liberals. "I cannot think of one that is so scandalous or so mischievous as, for instance, the call for nuclear disarmament, or world government, or what have you, upon which the press at large has yet to register its anxiety or scorn." By picking on the Birchers, Buckley declared, the liberals betrayed their own insecurity. "The only thing many of these critics would dislike more than a conservative organization with vulnerabilities is a conservative organization without vulnerabilities."[6]

If Buckley wouldn't disown the Birchers, neither would Barry Goldwater. The Arizona senator became the darling of the new conservatives somewhat by default. The death of Robert Taft in 1953 had left Republican conservatives leaderless, unless one counted Californian William Knowland, whose personal problems, in any event, forced him off the national stage in 1958. Richard Nixon was tainted by too-long association with Eisenhower. Even those conservatives who didn't consider Eisenhower a communist disapproved of the Republican president's failure to try to roll back the New Deal and his efforts to move the Republican party toward the center of American politics. Goldwater was no Eisenhower Republican, but he labored under the double disadvantage of being from a lightly populated state far from the centers of political and financial power, and of having compiled a less than distinguished record in the Senate. On the other hand, since conservatives sought to distance themselves from the groups that had brought America to what they considered the sorry mess the country was currently in, being from Arizona and possessing a low profile could count in Goldwater's favor. He certainly looked presidential enough, and in the dawning age of television, looks mattered.

Goldwater's conservative credentials were unimpeachable. He placed them on display, a trifle prematurely as it turned out, in a 1960 manifesto called *The Conscience of a Conservative*. Had Goldwater bared his conscience even four years later, he could have exploited the social disruptions that formed a primary source of conservative discontent during the mid-1960s. As matters stood, the book focused on the anti–New Deal message that had figured centrally in conservative speeches and literature for a generation. Goldwater equated liberalism with radicalism, and tied both to incipient absolutism. He decried "welfarism," and branded graduated income taxes a form of socialism. He defended states' rights against federal encroachment, and asserted that Washington had no business forcing schools to integrate. He denounced the Supreme Court for concocting novel and foolishly unwarranted interpretations of the Constitution.

Forthright as Goldwater's comments on domestic matters were, his remarks on international affairs were even more direct. After summarizing his critique of circumstances at home, he continued, "And still the awful truth remains. We can establish the domestic conditions for maximizing freedom, along the lines I have indicated, yet become slaves. We can do this by losing the Cold War to the Soviet Union." Goldwater would have none of the talk of peaceful coexistence. "We are in clear and imminent danger of being overwhelmed by alien forces. We are confronted by a world revolutionary movement that possesses not only the will to dominate absolutely every square mile of the globe, but increasingly the capacity to do so: a military power that rivals our own, political warfare and propaganda skills that are superior to ours, an international fifth column that operates conspiratorially in the heart of our defenses, an ideology that imbues its adherents with a sense of historical mission; and all of these resources controlled by a ruthless despotism that brooks no deviation from the revolutionary course." Americans failed to appreciate the full impact of this truth. While the Soviet Union was waging war against America, American leaders were—here Goldwater turned Eisenhower's favorite phrase against him—"waging peace."

Goldwater didn't object to peace, in principle. But he declared that peace must not be the primary objective of American policy. Vanquishing communism came first. "A tolerable peace, in other words, must follow victory over Communism. We have been fourteen years trying to bury that unpleasant fact. It cannot be buried, and any foreign policy that ignores it will lead to our extinction as a nation." Goldwater criticized containment as overly cautious and defensive. "Like the boxer who refuses to throw a punch, the defense-bound nation will be cut down sooner or later. As long as every encounter with the enemy is fought on his initiative, on grounds of his choosing and with weapons of his choosing, we shall keep on losing the Cold War." Goldwater charged that America's foreign-aid policy rewarded neutralism and strengthened socialism in the receiving countries. He lashed the United Nations, which, Americans needed to remember, was "in part a Communist organization."

The United States must seize the initiative in the Cold War. "Our strategy must be primarily offensive in nature." The American military must achieve and maintain superiority over the Soviets. "Mere parity will not do." The American government should proclaim the world communist movement to be "an outlaw in the community of civilized nations." To give meaning to this proclamation, the United States should sever diplomatic ties with the Soviet Union and all other communist countries. The United States should foment rebellion in Eastern Europe. "We should encourage the captive peoples to revolt against their Communist rulers." Likewise with respect to China, Korea, and Vietnam. "We should encourage friendly people that have the means and desire to do so to undertake offensive operations for the recovery of their homelands." Americans must get ready to enter the fray directly. "We must ourselves be prepared to undertake military operations against vulnerable Communist regimes." If another situation like that in Hungary in 1956 developed, the United States should issue an ultimatum to Moscow not to intervene. If Moscow refused the ultimatum, Washington should send military forces armed with nuclear weapons to the scene to compel compliance. Though war might result, a Soviet retreat was more likely.

"This is hard counsel," Goldwater granted. But it was necessary. America's future was in one of two directions. "Either the Communists will retain the offensive; will lay down one challenge after another; will invite us in local crisis after local crisis to choose between all-out war and limited retreat; and will force us, ultimately, to surrender or accept war under the most disadvantageous circumstances. Or we will summon the will and the means for taking the initiative, and wage a war of attrition against them— and hope, thereby, to bring about the internal disintegration of the Communist empire." The former course ran the risk of war and led, in any case, to probable defeat. The latter also risked war, but it held out the promise of victory. Americans must choose.[7]

Americans did choose in 1964, among candidates if not quite so clearly among the candidates' policies. Those who belonged to the Republican party chose Goldwater. To a substantial extent, Goldwater's nomination owed to the political aptitude of his supporters, especially the new-conservative types who orchestrated the draft-Goldwater movement. Yet if the Goldwaterites outmaneuvered the centrist supporters of Nelson Rockefeller, they also touched a strain in Republican thinking that would eventually transform the party and, to a somewhat lesser extent, the country.

Unfortunately for Goldwater, in 1964 the conservative wave hadn't formed sufficiently to carry his candidacy to victory. In fact, the nomination proved a fiasco. The nominee's acceptance declaration that "extremism in the defense of liberty is no vice," while hardly more provocative than John Kennedy's pay-any-price promises of anti-communist steadfastness, evoked shudders in Americans hardly twenty months after the Cuban missile crisis. Further, in the wake of the Kennedy assassination, even the canniest Republican would have had a hard time ousting the martyr's heir. Johnson lost no opportunity to play the "let us continue" theme, nor to portray his opponent as untested and irresponsible. By means of the Gulf of Tonkin resolution—a response to an event that on subsequent closer examination appeared never to have happened—Johnson preempted congressional and most editorial and public criticism of his handling of the Vietnam War. The measure of the incumbent's success in blunting Goldwater's challenge was the 16-million-vote margin by which Johnson buried the Republican nominee in the November election.

Smart money predicted that Republican conservatives would fold their tents and slip off into the Arizona desert and other lonely spots following the debacle. Many of them did. But a few wandered across the border into California to get a closer look at the one person who came out of the campaign with reputation enhanced. Ronald Reagan, for some time a booster of free-enterprise employed by General Electric, in 1964 co-chaired the California chapter of Citizens for Goldwater. Reagan taped a speech promoting Goldwater and conservatism that proved such a hit locally that the Goldwater campaign aired it nationally a week before the election. Although the speech didn't save Goldwater, who by this point was beyond resuscitation, it brought the telegenic and amiable Reagan to a political audience far wider than any he had addressed previously. Several years would pass before he landed the leading role in American politics, but on the basis of this initial screen test, his prospects looked good.

Reagan aside, the Goldwater candidacy served America principally as an anchor to starboard, a device that prevented American politics from drifting too far to the left. For all Johnson's claim of a mandate to honor the Kennedy legacy, neither the president nor anyone else could deny the fact that, despite Goldwater's liabilities and Johnson's advantages, the Republican candidate garnered 27 million votes. To win, Johnson had merely had to position himself to the left of Goldwater, which he did. But Goldwater, by positioning *himself* so far to the right, kept Johnson well to the right of where the Democratic candidate might otherwise have been.

If nothing else, the 1964 election demonstrated that the Cold War was alive and kicking. Though the voters weren't exactly demanding the militant liberationism Goldwater advocated, the support Goldwater received indicated that neither were they backing away from staunch opposition to communism. Goldwater didn't convince a majority that America was losing the Cold War, but he made it appear that defeat was an ever-present possibility.

The anti-antiwar movement

Though Johnson managed to keep Vietnam from becoming the dominant issue in the 1964 campaign, the war in Southeast Asia soon occupied a commanding position on the American scene. The president stuck to his middle-of-the-road strategy, refusing both the demands of the right for a declaration of war and general mobilization and the demands of the left for negotiations and withdrawal. But the strategy served only to prolong the war, and Johnson discovered that merely averting defeat required an ever increasing American commitment. Within a month of his 1965 inauguration, he approved a sustained bombing offensive against North Vietnam. Six months later, he authorized the dispatch of 100,000 additional American troops to Vietnam, although he initially announced the departure of only 50,000.

From the standpoint of provoking controversy, as well as of not winning the war, the bombing and the troop increase proved about equally effective. The bombing had the advantage, in Johnson's view, of involving small numbers of Americans and producing few American casualties. It utilized America's superiority in firepower, and took the war to the enemy's home turf. But it had the unavoidable disadvantage of casting the United States as a horrible bully, a superpower pounding a peasant country to dust for opposing the American will. American leaders described the bombing as defensive, and by their lights it was. The United States was bombing North Vietnam to defend South Vietnam. Yet the logical connection seemed tenuous to many who thought that, since the fighting was taking place in South Vietnam, the defending ought to be taking place in South Vietnam as well. Further, although American warplanes attempted to restrict their attacks to military targets, the bombing inevitably killed and injured many civilians. Because the official American position was that the government of North Vietnam was a dictatorship, unresponsive to the wishes of the North Vietnamese people, it seemed particularly perverse for the United States to punish those people for the sins of their government.

Johnson's decision on troops, and his method of revealing the decision, likewise turned out to have a double edge. By declining to declare a national emergency, to call up the reserves, or otherwise to put the country on a war footing, Johnson mitigated the political impact of the escalation. Moreover, by announcing the increase of troops in two stages, he made the escalation appear less drastic than it was. But by putting so many Americans on the front lines, he inescapably increased American casualties, thereby guaranteeing a certain level of discontent, especially when the danger to the United States was indirect and largely theoretical. In addition, the policy of incrementalism placed the Johnson administration in the position of seeming to play less than straight with the American people. The term "credibility gap" had already entered the language, largely as a result of Johnson's confused and misleading attempts to justify the April 1965 invasion of the Dominican Republic. The press had knocked holes in the president's fevered assertions that Castroite communists stood on the verge of converting the Dominican Republic into a second Cuba, and by the end of the summer, even J. William Fulbright, the Democratic chairman of the Senate foreign relations committee, had gone into opposition. Johnson, who entered the White House with a reputation as a Texas wheeler-dealer, would have had to watch his step in any event. But the Dominican affair, followed shortly by the rationing of the Vietnam troop announcement, simply confirmed the opinion of many observers that the president was trying to pull some fast ones on the American people—which he was.

Consequently, Johnson had himself, as much as his bad luck, to blame for the rapid emergence of the strongest antiwar movement America had seen since the Mexican War. (The Civil War, in which half the country opposed the war—by means of war—was a special case.) Anyone might have guessed that the movement would strike sparks on the campuses, the predictable hotbeds of radicalism. Had the movement been confined to the colleges and universities, it might have been ignored. But by the time the protests reached those last and nearly always reliable redoubts of prowar sentiment, veterans' organizations, it was clear that the opposition was really serious. (By then, though, Johnson was out of office, and the problem was Richard Nixon's.)

Initially, the Johnson administration treated the opposition to the war as chiefly the result of a failure to get the administration's point across to the American people. "We have an education problem that bears watching and more work," McGeorge Bundy remarked to Johnson early in 1965. To remedy the problem, the administration attempted to restate its arguments in more forceful fashion. In news conferences, at photo opportunities, and in full-dress speeches, the president recited the central themes of the Cold War. Communism was trying to expand, Johnson said, and it would succeed in doing so unless stopped by the United States. The world was watching to see if Americans possessed the will and stamina to halt the aggressors. If the United States failed to stand firm, the American alliance system would disintegrate, and countries from Germany to Japan would seek accommodation with the Kremlin. The free nations had failed to stand up to aggression at Munich in 1938, and the worst war in history had followed. Irresolution in Vietnam in the 1960s could prove as disastrous— or worse, in the age of nuclear weapons.[8]

Other administration officials joined the effort to demonstrate that defending countries like Vietnam was what the Cold War was all about. As part of a nationally televised "teach-in" on the war, McGeorge Bundy debated arch-realist Hans Morgenthau, and explained that the administration's policies in Vietnam demonstrated nothing so much as a hard-headed assessment of American interests. (Johnson didn't like the idea of his national security adviser lending stature to the opposition by participating in such a debate, but Bundy felt a responsibility to his own intellectual integrity, and that of the administration, to take part.) Bundy and Dean Rusk appeared on television interview shows and before congressional committees to justify the president's actions as being entirely consistent with a generation of Cold War policy. The White House put together a "truth squad" that visited campuses in the wake of antiwar speakers, to present the case for the war.

Johnson meanwhile leaned on friends, former colleagues, and anyone else he could think of to back organizations and activities that supported his approach to the troubles in Southeast Asia. He called on officials of former administrations—Dean Acheson, Dwight Eisenhower, and others—to corroborate his assertions that what he was doing in Vietnam was just what they would have done. Though Eisenhower, for one, didn't entirely approve of Johnson's management of the war—Eisenhower thought that if the United States was going to fight, it should fight to win, and win soon—the old general was a good enough soldier not to differ publicly with the commander-in-chief. Johnson knew he wouldn't.

Even as he defended the war on its merits, Johnson sought to discredit opponents of the war. Some critics gained Johnson's grudging respect. The president had little bad to say about Wayne Morse and Ernest Gruening, the only two legislators to vote against the 1964 Tonkin Gulf resolution. They spoke their minds forthrightly, and held to a steady point of view. Besides, they didn't occupy particularly influential positions. But for William Fulbright, Johnson had considerably less regard. The president thought that Fulbright, as the most powerful Democrat in the Senate on foreign affairs, owed more to the party than he was giving. Further, unlike Morse and Gruening, Fulbright changed his mind on Vietnam after the escalation began. Johnson took this as indicating a lack of fortitude and courage, if not downright duplicity. Finally, Johnson believed that jealousy and personal pique motivated much of Fulbright's opposition. Johnson told a reporter, off the record, "Fulbright has never found any president who didn't appoint him secretary of state to be satisfactory."[9]

When Fulbright held hearings on the war during the spring of 1966, Johnson did his best to prevent the hearings from gaining widespread attention. Fulbright found certain administration witnesses hard to locate or unwilling to testify, and he later charged that the president had deliberately sent some of them out of the country and had muzzled the others. When it became clear that the hearings nevertheless were generating considerable interest, to the point that the major television networks preempted the afternoon soap operas to carry the testimony live, Johnson scheduled a snap summit meeting with South Vietnamese prime minister Nguyen Cao Ky in Honolulu. The effect was to require the media to pull cameras and reporters away from Fulbright and the opposition, and direct them toward Johnson and the war effort.[10]

Johnson didn't like Fulbright, and he distrusted the Arkansas senator's motives and judgment, but at least he didn't question Fulbright's loyalty to the United States. The same couldn't be said of Johnson's opinion regarding the more extreme of the antiwar protesters. Not surprisingly, and not entirely unreasonably, Johnson suspected that at least some of the demonstrations against the war reflected the influence of communists in the United States, and of persons generally who felt a stronger allegiance to Moscow or Beijing than to Washington. In April 1965, Johnson met with J. Edgar Hoover, the director of the Federal Bureau of Investigation, to discuss his concerns. As Hoover related the conversation to aides afterward, the president said he had "no doubt" that communists were "behind the disturbances." Hoover agreed, asserting that radicals affiliated with the Students for a Democratic Society were planning protests in scores of cities, and that the SDS was "largely infiltrated by communists."[11]

Johnson ordered Hoover to explore the connection between communists and the antiwar movement. Hoover put the FBI on the trail at once. The director told his subordinates that he wished to see "a good, strong memorandum" showing how communists had been participating in demonstrations. Hoover admitted to his aides that the bureau couldn't prove that communists had initiated all the antiwar activities, which they obviously hadn't. But the bureau could highlight the facilitative role the communists had been playing. "What I want to get to the President is the background with emphasis upon the communist influence therein, so that he will know exactly what the picture is."[12]

Hoover, and then Johnson, got the memo the director ordered. It delineated attempts by the American Communist party to encourage domestic dissent by "a crescendo of criticism aimed at negating every effort of the United States to prevent Vietnam from being engulfed by communist aggressors." While recognizing that most groups involved in the antiwar movement weren't controlled by the communists, the memo contended that the party had "vigorously supported these groups and exerted influence."[13]

Hoover had been hunting subversives since the first Red Scare of the post–World War I era, and he hardly needed Johnson's encouragement to keep an eye on radicals and other boat-rockers. From the early years of the Cold War, the FBI had infiltrated and attempted to disrupt organizations Hoover saw as posing an unacceptable risk to the American government or the American status quo. During the 1950s, the FBI employed a large number of informants inside the American Communist party. One cause for the ease with which the FBI penetrated the party may have been that the party, then in a moribund state, was in no position to turn down applicants for membership. Anti-communist prosecutions under the 1940 Smith Act had resulted in the imprisonment of much of the party's leadership, and the McCarthyist investigations had scared off many of the less committed. Khrushchev's 1956 revelation of the brutality and megalomania of Stalin, who for three decades had been the hero and guiding light of world communism, hit remaining loyalists hard, as did the Soviet invasion of Hungary, which demonstrated that the homeland of socialism could engage in imperialism as oppressive as that of the West. (Eisenhower's strong stance against the simultaneous British-French-Israeli attack on Egypt made Moscow's treatment of the Hungarians look that much worse.) By the beginning of 1957, the entire membership of the American Communist party could have fit inside a high-school gymnasium. Such a large portion of the membership worked for the FBI that at one point Hoover considered a political takeover of the party, to be accomplished by throwing the votes of all the FBI's people to a single, manipulable faction in one of the party's innumerable splits.[14]

During the Kennedy years, other radical groups also attracted FBI attention. In 1961, Hoover sent a memo to FBI district offices ordering them to devise schemes for disrupting the Socialist Workers' Party. Hoover defended the disruption on grounds that the SWP had been promoting revolution by "strongly directing and/or supporting such causes as Castro's Cuba and integration problems arising in the South." Referring to an ongoing program of subversion against the Communist party, Hoover wrote, "It is felt that a disruption program along similar lines could be initiated against the SWP on a very selective basis." Hoover wanted the bureau to pick its spots cautiously. "This program is not intended to be a 'crash' program. Only carefully thought-out operations with the widest possible effect and benefit to the nation should be submitted."[15]

In response to Hoover's memo, bureau headquarters received numerous recommendations from the field. The Detroit office was especially helpful. Detroit suggested sending letters signed by fictitious individuals to newspaper editors, describing the radical antecedents of SWP candidates for political offices. Such letters should make clear the revolutionary aims of the SWP. "This procedure would alert the public to the fact that in voting for a SWP candidate they are not voting for an innocuous 'Socialist' candidate or a 'labor' candidate, but for a candidate dedicated to the overthrow of the United States Government." The bureau might surreptitiously produce leaflets containing similar information. "The information could be forwarded to newspaper reporters, radio and television stations, always taking necessary precautions so that the identity of the FBI as the source would not be disclosed." The bureau could send letters to supporters of organizations in which SWP members played important roles, apprising the supporters of the "subversive connections" of the SWP members. Landlords who rented meeting halls to SWP-infiltrated organizations could be told of the subversive uses to which their premises were being put. "This could be repeated, causing the disruption of proposed meetings."[16]

Available evidence doesn't indicate whether Hoover approved these recommendations, or if he ordered implementation of such other suggestions as one from the Denver office that letters from "A Concerned Mother" be sent to newspapers in the Denver area underlining the SWP affiliation of the husband of a candidate for the Denver school board, and noting the fact that the United States attorney general had included the SWP on his list of subversive organizations. Hoover did, however, approve the sending—with the "usual precaution taken in order that the letter cannot be traced back to the Bureau"—of a spurious letter designed to disrupt the 1969 candidacy for New York mayor of Paul Boutelle. The purpose, according to the New York office, was "to further polarize blacks and whites in the SWP." The letter ostensibly was written by several white SWP members, who told Boutelle, an African-American: "Some of us within the Party are fed up with the subversive effects you are having on the Party, but since a few see your presence as an asset (because of your color only) not much can be said openly." The letter described Boutelle as "utterly useless to the revolution" and concluded, "Why don't you and the rest of your fellow party monkeys hook up with the Panthers where you'd feel at home?"[17]

By this time, the FBI had received reinforcements in its anti-antiwar campaign. In 1967, the CIA signed on, launching Operation CHAOS, which began as an offshoot of the agency's overseas counterintelligence program, but soon developed a life of its own. In trying to determine the extent to which Moscow, Beijing, and Havana were, in the words of the operation's initial instructions, "exploiting our domestic problems in terms of espionage and subversion," the CIA conducted investigations of thousands of Americans of liberal-to-radical political views. Not only were the methods the agency used—opening mail, for instance—legally and ethically questionable, but the entire operation violated the agency's charter, which explicitly forbade it from engaging in domestic intelligence. (The forbidding clause had been designed only partly to protect Americans from the development of a Gestapo. It had also served to buy the acquiescence of Hoover, who bull-doggedly guarded his fiefdom against bureaucratic encroachment.) CIA officials were fully aware of what they were doing. Referring to a study of America's "restless youth," director Richard Helms commented, "This is an area not within the charter of this Agency, so I need not emphasize how extremely sensitive this makes the paper. Should anyone learn of its existence, it would prove most embarrassing for all concerned."[18]

Civil wrongs

While storing up embarrassment, officials of the American government directed their anti-radical attention to another area where challengers to the status quo were making noises: race relations. Allegations of links between communism and persons advocating racial equality were nothing new. Since at least the 1930s, opponents of civil rights reform had claimed that communists lay behind the agitation for racial equality. Before the onset of the Cold War, such claims carried only modest penalties. Reds ran in the best crowds in the days of the anti-fascist popular front, and after a brief moment of obloquy following the 1939 Molotov-Ribbentrop pact, they re-emerged as worthy allies. By the 1950s, however, charges of communist connections were taken quite seriously. Hoover's FBI certainly took the charges seriously, and when reports surfaced of ties between the civil rights movement and the American Communist party, the bureau put its trackers to work. Officially, the bureau's watchfulness indicated a concern that American communists might try to take over the civil rights movement, a possibility American leaders and the American public should be made aware of. In reality, the bureau's—and particularly Hoover's—intense interest in any links between communists and the black movement reflected an ideological aversion to anything that shook the status quo, as well as what seems to have been a tendency toward voyeurism regarding the personal affairs of African-American leaders, especially Martin Luther King, Jr.

There was just enough truth to allegations of a communist role in the drive for racial equality to give a gloss of respectability to the bureau's preoccupation. The position of the American Communist party on race was a curious, but not illogical, amalgam of theoretical disdain for the revisionist illusion that desegregation, rather than proletarian revolution, would solve the problems of American blacks, and of everyday egalitarianism in organizational and political matters. It was a commonplace that federal agents working in the South spotted white communists by the fact that they associated easily with African-Americans at a time and under circumstances when and where almost no other whites did.

Not surprisingly, many people who objected to government persecution of American communists also objected to discrimination against American blacks. Among the most influential of those sympathetic to each group was Stanley Levison. During the early 1950s, Levison had contributed to defense funds for individuals prosecuted for their communist beliefs, and he had helped manage the finances of the Communist party—perhaps including covert subventions from the Soviet Union. After the Montgomery bus boycott began in 1955, however, Levison's interests shifted to civil rights. He worked to establish a group called In Friendship, which supported Southern blacks against segregationist violence and which in 1956 sponsored a large rally at Madison Square Garden, with the proceeds going to the Montgomery boycotters. Levison met Martin Luther King about this time, and soon became King's closest white confidant and adviser.

The connection between those who supported civil liberties for communists and those who supported civil rights for blacks wasn't lost on opponents of change in America's racial hierarchy. Expatiating on the significance of the connection became a staple of conservative editorialists, especially in the South, but elsewhere also. The *Columbia Record* of South Carolina in May 1964 noted a recent account by "the liberal columnist Joseph Alsop" linking King to certain communist influences. Publication of this account, the *Record* said, was followed by King's unexplained absence from a convention of American newspaper editors. "That King associated with Communists, attended Communist front meetings, and had suspect individuals in key positions in his organization was no secret," the Columbia paper asserted. "Facts had been disclosed, but they were revealed by conservative elements and therefore were ignored and discarded. Alsop's column, plus King's disappearance from his customary high degree of public exposure, indicated Communist influence of surpassing significance." The *Charleston News and Courier,* foreseeing a summer of demonstrations and possibly violence in 1964, declared, "Never invaded by a foreign power since the War of 1812, the United States will be fighting the war of communism on its own soil this summer. . . . This country should be warned, and should be braced, for a Negro insurrection which is nothing less than a Communist invasion of American cities." The *Washington Evening Star* remarked on a "new militancy" in the civil rights movement. "Some new groups have been formed which seem more inclined to violence or overt mass acts which might lead to violence. As a result of this, some observers who a year ago would have scoffed at charges of Communist influence now are raising warnings about it." *U.S. News and World Report*, recounting violence in various cities, declared, "All the disgraceful episodes which have occurred in New York and other cities recently were certainly not directed by patriotic American Negro leaders. . . . The time has come for the Government of the United States to do more to expose the infiltration in civic movements by the Communist Party and its agents, stooges, and allies inside this country." The *Washington Post* carried a column by Rowland Evans and Robert Novak describing demonstrations in California and commenting, "Here as elsewhere the Negro establishment is in danger of losing control over the civil rights movement to thugs and Communists."[19]

The theme of a Red-black conspiracy played especially well in the South's fundamentalist religious press. The Christian Crusade of Tulsa produced a typical tract under the title *Unmasking the Deceiver.* The author, Dr. Billy James Harris, asserted, "Recent statements by the race agitator Martin Luther King, Jr., clearly indicate that it is time to rip off his pious mask and reveal the real purpose and drive behind his anti-American activities. Though King has been sainted in many popular weekly magazines, his infamous alliance with Communist objectives and personalities has been a carefully guarded secret." Harris judged that duty required him to clue in his readers to this secret. He did so at some length, summarizing his case with the statement: "King's Communist affiliations and acquaintances go back many years and establish a clear pattern of Marxist affinity."[20]

White Southern politicians naturally got into the act. William Dickinson of Alabama took the floor of the House of Representatives to relate in detail the peace-disturbing activities of various segments of the civil rights movement. Dickinson proceeded to ask, "Who then is the one or group that puts these groups together—that gives them cohesiveness, strength, money and direction? Who or what can weld this diverse group together into a formidable force that can—and has—overcome? The answer is this: the Communist Party." What of the leader of the movement? "Martin Luther King himself has amassed the staggering total of more than 60 Communist-front affiliations since 1955."[21]

Senator Strom Thurmond of South Carolina lodged a like complaint in hearings on what became the Civil Rights Act of 1964. Thurmond didn't deny the sincerity of much of the feeling behind the parades and other demonstrations calling for an end to segregation. But he requested that Congress consider where the inspiration for all the agitation came from.

"These parades, in my judgment, are inspired by the Communists, and I think it is part of the international conspiracy of communism."[22]

To these same hearings, Southern governors brought similar views. George Wallace of Alabama stated unequivocally, "There are Communist influences in the integration movement." Wallace identified "a known Communist," Jack O'Dell, as one of King's close entourage. Wallace reported that King had "publicly professed" to have fired O'Dell from the Southern Christian Leadership Conference, but in fact had failed to do so. "The public profession was a lie." O'Dell remained on the SCLC payroll. (Not for long. King, under pressure from friends of the civil rights movement as well as from enemies, severed O'Dell shortly after this.) Wallace went on to cite a picture of King and "a group of Communist and pro-Communist leaders" attending a meeting together. Among those shown was Fred Shuttlesworth, "a self-styled 'reverend' " and president of the Southern Conference Educational Fund. Wallace reminded his listeners that investigators from both the Senate and the House of Representatives had identified the SCEF as an organization "set up to promote communism." Shuttlesworth had condemned himself out of his own mouth by challenging the good faith of Congress and of the people of the South. Wallace read an newspaper article quoting Shuttlesworth: "Generally, the House committees are governed by southerners who will label any organization subversive or communistic that seeks to further the American aims of integration, justice, and fair play. To a segregationist, integration means communism." As further evidence, Wallace pointed out that Algerian prime minister Ahmed Ben Bella—"a Communist, in my opinion"—while on a visit to America had met with Martin Luther King. "And then Ben Bella flew to Cuba and embraced the Communist Castro and said that he is one of the world's greatest. Is there any connection?"[23]

Mississippi governor Ross Barnett thought so. Barnett spied a characteristic pattern in the activities of integrationists in the United States and of radical revolutionaries abroad. "What is happening in our nation today fits the pattern of what has been happening throughout the world insofar as the Communist activity is concerned," Barnett asserted. "Communist tactics are to create a crisis and then leave the scene with heartaches, turmoil and strife." Communists had employed these tactics in Cuba, Laos, Vietnam, Berlin, and elsewhere. Mentioning the American communities of Birmingham, Jackson, and Danville by name, Barnett said, "The same tactics are being practiced in the United States." The governor suggested that the communists were agitating the race question in order to get America to drop its guard. "Perhaps this is all part of a great conspiracy to divert our attention to this domestic issue so that we may neglect other and far more important matters." Whatever the tactical design, Barnett had no doubt as to the guiding strategy. The communists were "championing the cause of the Negroes in America as an important part of their drive to mobilize both colored and white for the overthrow of our government."

The evidence was plain. "I am thoroughly convinced that this is a part of the world Communist conspiracy to divide and conquer our country from within."[24]

Barnett, Wallace, and others thought that federal authorities ought to quit forcing integration down the throats of law-abiding citizens, and concentrate instead on investigating the links between black leaders like King and the communists. Barnett and the others were evidently unaware that the FBI already had a long file on King and was adding to it daily. In August 1963, just before the historic civil rights march on Washington, the FBI's domestic intelligence office provided Hoover a sixty-eight-page report tracing efforts by members of the American Communist party to turn the civil rights issue to communist account. Although the report asserted that the efforts had failed, Hoover refused to accept this conclusion. Across the cover of the report he wrote, "This memo reminds me vividly of those I received when Castro took over Cuba. You contended then that Castro and his cohorts were not Communists and not influenced by Communists." Time had proved this estimate wrong, Hoover said. (Actually the estimate was essentially right: Castro had embraced communism only after assuming power. But Hoover doubtless considered this to be hairsplitting.) Hoover indicated that time would prove claims of only minor communist influence in the civil rights movement wrong as well.[25]

Hoover's subordinates generally jumped at the slightest sign of the director's displeasure, and they jumped on this occasion. In a mea culpa that sounded like something from Moscow during the Stalinist show trials, or Beijing during the Cultural Revolution, FBI assistant director William Sullivan chided himself and his associates for their obtuseness. "The Director is correct," Sullivan confessed. "We were completely wrong about believing the evidence was not sufficient to determine some years ago that Fidel Castro was a communist or under communist influence. On investigating and writing about communism and the American Negro, we had better remember this and profit by the lesson it should teach us." Just the day before Sullivan wrote this letter, King had delivered his "I have a dream" address. The performance impressed Sullivan, in a negative way. "Personally, I believe, in the light of King's powerful demagogic speech yesterday, he stands head and shoulders over all the other Negro leaders put together when it comes to influencing great masses of Negroes. We must mark him now, if we have not done so before, as the most dangerous Negro of the future in this Nation from the standpoint of communism, the Negro, and national security." Sullivan went on to assert, "Nineteen million Negroes constitute the greatest single racial target of the Communist Party, U.S.A. This is a sombre reality we must never lose sight of."[26]

Hoover appropriated Sullivan's phrasing almost word for word in testimony before Congress in the spring of 1964. "The approximate 20 million Negroes in the United States today"—Hoover inflated Sullivan's figure for heightened effect—"constitute the largest and most important racial target of the Communist Party, U.S.A." Hoover added, "The infiltration, exploitation and control of the Negro population has long been a party goal and is one of the principal goals today." Hoover didn't claim that all African-American leaders were communists or dupes. Yet a small cadre could suffice to capture the entire movement. "Communist influence does exist in the Negro movement, and it is this influence which is vitally important. It can be the means through which the large masses are caused to lose perspective on the issues involved and without realizing it succumb to the party's propaganda lures."[27]

Even as Hoover testified, the FBI was busy trying to smear King. At least since early November 1963, bureau personnel—with the approval of Attorney General Robert Kennedy, who was worried not only about what Hoover might find on King, but also about what Hoover had found on the Kennedys—had been listening via secret wiretaps to King's telephone conversations from his home and from SCLC headquarters in Atlanta. The wiretaps contributed data to the revised estimate of the King-communist connection that Hoover had demanded. This version traced the close relationship between Levison, a "dedicated Communist," and King. The relationship afforded the communists a crucial opening, the report said, which they intended to exploit. "The current atmosphere in the Communist party is marked by a vigorous spirit of enthusiastic optimism and a determination to launch more open, aggressive action on the national scene. As the situation now stands, Martin Luther King is growing in stature daily and is the leader among leaders of the Negro movement. Communist party officials visualize the possibility of creating a situation whereby it could be said that, as the Communist party goes, so goes Martin Luther King, and so also goes the Negro movement in the United States."[28]

The FBI didn't intend to let events reach such a pass. In December 1963, bureau officials met in Washington to conduct what Sullivan described as "a complete analysis of the avenues of approach aimed at neutralizing King as an effective Negro leader and developing evidence concerning King's continued dependence on communists for guidance and direction." The wiretaps and additional intelligence gathered from FBI informants suggested one promising lead. "Although King is a minister, we have already developed information concerning weaknesses in his character which are of such a nature as to make him unfit to serve as a minister of the gospel." The bureau officials at this meeting decided on a six-point program:

(1) We must determine and check out all of the employees of the SCLC.
(2) We must locate and monitor the funds of the SCLC.
(3) We must identify and check out the sources who contribute to the SCLC.
(4) We must continue to keep close watch on King's personal activities.
(5) We will, at the proper time when it can be done without embarrassment to the Bureau, expose King as an immoral opportunist who is not a sincere person but is exploiting the racial situation for personal gain.

(6) We will explore the possibility of utilizing additional specialized investiga-
tive techniques at the SCLC office.

Sullivan's notes of the meeting closed with a reiteration and amplification of
the fifth point: "We will continue to give this case priority attention both at
the Seat of Government [Washington] and in the field and will expose King
for the clerical fraud and Marxist he is at the first opportunity."[29]

The FBI's harassment campaign against King took a variety of forms.
Bureau personnel compiled an edited version of some of the wiretapped
recordings it had on King, and sent these to him anonymously as a warning. A likewise anonymous letter accompanied the tape. "King, look into
your heart," the letter commanded. "You know you are a complete fraud
and a great liability to all of us Negroes. White people in this country have
enough frauds of their own, but I am sure they don't have one at this time
that is anywhere near your equal." The letter continued, "King, like all
frauds your end is approaching. . . . There is only one thing left for you to
do. You know what it is." The one thing presumably was suicide, or at
least retirement from the civil rights movement. The letter concluded,
"You are done. There is but one way out for you. You better take it before
your filthy, abnormal, fraudulent self is bared to the nation."

Beyond writing poison-pen letters, the bureau bribed an accountant at
SCLC headquarters to provide tidbits not supplied by the planted micro-
phones, and it prevailed upon the Internal Revenue Service to examine
King's back tax returns, looking for evidence of useful irregularities. When a
small African-American newspaper commented unfavorably on King, the
bureau discreetly reproduced the editorial for wider distribution. The bureau pulled together the juiciest of its information about King and served the
goodies up to the president and other top government officials.[30]

As the civil rights movement broadened during the mid-1960s, the FBI
cast its net more widely. The bureau watched with particular concern the
growing appeal of militant black nationalist organizations such as the Nation of Islam, or Black Muslims. Without forgetting King and the SCLC,
the bureau developed a program for subverting the militants. In a memo of
March 1968, the bureau's counterintelligence division listed priorities for
dealing with what it called "Black Nationalist-Hate Groups." Topping the
list was heading off a possible coalition among different groups. "An
effective coalition of black nationalist groups," the memo explained,
"might be the first step toward a real 'Mau Mau' in America, the beginning
of a true black revolution." Further, the bureau should work to keep the
black nationalists from gaining respectability among blacks and whites
both, and it should especially strive to block the growth of black nationalist sentiment among black youth. Finally, the bureau should seek to "prevent the rise of a 'messiah' who could unify, and electrify, the militant
black nationalist movement."

The March 1968 memo went on to say:

Malcolm X might have been such a "messiah"; he is the martyr of the movement today. Martin Luther King, Stokely Carmichael, and Elijah Muhammed all aspire to this position. Elijah Muhammed is less of a threat because of his age. King could be a real contender for this position should he abandon his supposed "obedience" to "white, liberal doctrines" (nonviolence) and embrace black nationalism.[31]

When this memo became public several years later, it raised questions regarding possible involvement by the FBI in King's assassination, which occurred just weeks after the memo was written. Hoover's general animus toward King had been well known during the 1960s. In 1964, after King charged FBI agents with sympathizing with Southern whites involved in violence against blacks, Hoover openly branded King "the most notorious liar in the country." King and his associates were aware that Hoover was using the resources of the FBI to cause them trouble. Not unnaturally, therefore, some among King's survivors suspected that the bureau may have played a part, perhaps indirect, in prompting James Earl Ray to shoot King. The release of the 1968 memo did nothing to diminish these suspicions, although the evidence it afforded was merely circumstantial.[32]

Whatever the anti-King campaign reveals about Hoover and the FBI, it demonstrates again the degree to which the Cold War had thoroughly infiltrated American life by the 1960s. It also demonstrates again the extent to which the Cold War had been turned to purposes essentially unrelated to international affairs. For those who opposed integration and other forms of racial amelioration, the alleged communist threat provided a means of avoiding the real issues in the civil rights debate. Were African-Americans being treated unfairly? Was segregation unjust? By raising questions about King's ties to communists, segregationists such as Wallace and Barnett and Thurmond and less-focused reactionaries such as Hoover shifted the terms of the argument. Agitation by blacks became a threat not merely to Southern institutions but to the American way of life.

The segregationists ultimately lost the argument, of course, although their rearguard action delayed and mitigated their defeat. The defeat resulted primarily from the moral irrefutability of black claims on the American conscience. Cold War pseudo-patriotism, however earnestly invoked, couldn't forestall these claims forever. Yet for a considerable while it did, providing a last refuge for racists and other scoundrels.

Liberalism lost

A worse year for America than 1968 would be hard to imagine. Beginning in January, the North Vietnamese and Viet Cong launched the Tet offensive, which shattered what remained of the illusion that victory in Vietnam was anywhere in sight. By the middle of February, such pillars of the American establishment, and erstwhile supporters of the war, as Dean

Acheson, Walter Cronkite, and the *Wall Street Journal* were calling for a negotiated settlement. Johnson's credibility gap became a yawning chasm that swallowed the president, who in March announced his withdrawal from the 1968 election campaign. Less than a week after Johnson's announcement, King was assassinated. The King killing stunned those who had hoped to witness the peaceful revolution he advocated, and it triggered a wave of urban violence surpassing anything the country had yet seen—which by 1968 was quite a lot. In the aftermath of King's death, black despair and outrage ignited more than sixty cities, producing scores of deaths, thousands of injuries, tens of thousands of arrests, and hundreds of millions of dollars of property damage. Before the flames died away, another assassination, this of Robert Kennedy, threw America into further disarray. If King's death seemed to indicate that the system wouldn't brook even nonviolent demands from outside for change, Kennedy's appeared to demonstrate that change from within the system was equally impossible.

The summer of 1968 brought no relief. When the Democrats met in Chicago to nominate a person they hoped would succeed Johnson as president, the nation was confronted with the televised spectacle of an army of police bashing the heads of demonstrators, news reporters, and careless onlookers. Those watching had to decide who posed a greater threat to the republic: the Yippies and the others who appeared to go out of their way to flaunt their contempt for order, reason, and the rest of the value system of a civilized society, or the police, who seemed determined to prove that everything the Yippies said about fascism taking over America was true.

The Chicago riots climaxed the chaos of the previous several years, and produced a revulsion among voters that helped drive the Democrats from the White House, sending them into an exile relieved only once until the Cold War's end. The conservatives who had rallied behind Goldwater in 1964 might have capitalized on the distress but for two developments: the return of Richard Nixon and the breakaway of George Wallace. Nixon was cleverer than Goldwater had been in 1964. He positioned himself as a candidate who shared essential values with conservatives but wouldn't scare away moderates. The strategy drained conservative support from Ronald Reagan's candidacy, and gave Nixon the nomination. Meanwhile, George Wallace abandoned the Democratic party and mounted a campaign of protest against all the protests. Wallace attacked federal enforcement of civil rights in the states. He denounced busing to achieve integration. He blasted antiwar demonstrations. He praised intolerance of divergent lifestyles. Pointing to a group of protesters at one of his rallies, Wallace summarized his message: "These are the folks that people in this country are getting sick and tired of. You'd better have your day now because come November, I tell you, you are through."[33]

Nixon and Wallace split the conservative vote, almost allowing Hubert Humphrey to slip into the White House. Together, Nixon and Wallace took nearly 57 percent of the popular tally, to Humphrey's less than 43 percent. Though not all Wallace's ballots would have gone to Nixon in a two-candidate race, the result still represented a massacre of American liberalism. Regarding domestic affairs, this appeared to mean a repudiation of the idea that government should be the agent of first—or early, or maybe even last—resort in effecting change. Regarding foreign affairs, it signaled a reaffirmation of the anti-communist, anti-appeasement principles of the Cold War. Or seemed to.

★

CHAPTER FIVE

What Did We Know and When Did We Know It? 1969–1977

Secret plans, secret wars, secret secrets

It soon became clear that with Nixon, America could never be too sure what to expect. Nixon had been one of the original Cold Warriors—a relentless red-baiter in Congress, and a foreign-affairs hawk as Eisenhower's vice president. Of the three candidates in the 1968 contest, Nixon seemed the surest bet to stick to the philosophy that had guided American foreign policy since the late 1940s. Humphrey paid lip service to anti-communist containment, but lukewarmly and largely out of loyalty to Johnson. His heart had never been in the Cold War, and never would be. Wallace talked a decent anti-communist game, but the Alabama governor's brand of reactionary populism might easily turn isolationist. Although Nixon was less inspiring than either Humphrey or Wallace, he was experienced and steady. He offered more of the same, only better administered.

Regarding Vietnam, he offered specifically a "secret plan" to end the war. Nixon didn't invent the phrase, which originated with a reporter looking for a lead to a story summarizing the Republican candidate's nebulous promise to end the war without losing. But neither did he disavow the term, and it soon became a part of the campaign. When pressed for details, Nixon retreated to the not indefensible position that to tip his hand would interfere with the negotiations that had begun in Paris in May.

As implemented, Nixon's plan involved not so much ending the war as de-Americanizing it—which ironically (and so like Nixon, Humphrey thought) was the position the Democratic candidate had staked out during the campaign. Nixon was nothing if not shrewd, and he understood that most of the American opposition to the war resulted from the emotional and financial burden it placed on Americans. Once the body bags stopped coming back to the United States, and once the strain on American taxpayers eased, American disaffection would diminish. As it did, Nixon would gain the breathing space he needed to work out a solution to the Vietnam predicament.

Vietnamization, as the process was called, combined the withdrawal of American troops with demonstrations of continued American resolve to keep South Vietnam in the American camp. From more than half a million soldiers when Nixon took office in 1969, the American military presence in Vietnam steadily shrank until the last units departed at the beginning of 1973. Concurrently, the United States increased shipments of military aid to South Vietnam. By 1973, the South Vietnamese army was one of the three or four best-outfitted in the world. To reinforce the buildup, Washington ordered attacks on communist sanctuaries and supply lines across the border from South Vietnam in Cambodia and Laos. In 1969, American warplanes commenced a covert offensive in Cambodia, and in 1970 American troops followed the bombings with a ground assault. At the beginning of 1971, American planes provided air support for a major South Vietnamese operation in Laos.

This widening of the war, even as American involvement was growing more shallow, provoked greater domestic opposition than ever. Demonstrators closed campuses all across the country. At Kent State University in Ohio and Jackson State in Mississippi, violence turned fatal. Six demonstrators were killed—four at Kent State, two at Jackson State. For the first time in a generation, Congress reasserted its role in war-making, passing a measure forbidding future incursions by American troops into Cambodia. The *New York Times* began publishing excerpts from a purloined Pentagon study of the war, over the vigorous protests of the administration.

The leaking of the Pentagon Papers provoked the Nixon administration to actions that led to a major test of American constitutional government, to the president's downfall, and to fresh challenges to the Cold War premises of American foreign policy. The first step down this fateful path was innocuous enough: the administration sued to prevent the *Times* from publishing the Pentagon Papers. The White House was particularly sensitive to leaks at just this time, since administration negotiators were nearing the end of painstaking talks leading to a strategic arms limitation agreement, and the last thing the administration needed was for someone to reveal what Washington considered vital and what it considered dispensable in the SALT negotiations. But sure enough, in July 1971, the *Times* carried a story describing the administration's bottom line. Nonetheless, it was with some ambivalence that the administration sought an injunction against publication of the Pentagon Papers. The papers included almost nothing that genuinely required continued classification, and such chagrin as they caused would reflect chiefly on the Kennedy and Johnson administrations. (Nixon's chief of staff H. R. Haldeman insisted that his subordinates call them the "Kennedy/Johnson papers.") In fact, Haldeman thought the publication of the papers might redound to the Nixon administration's benefit. "We have nothing to hide," Haldeman declared. "This is a family quarrel in the previous Administration regarding the Kennedy/ Johnson conduct of the war. We, on the other hand, have developed a new policy, and it's working."[1]

All the same, the publication of classified materials violated federal law and threatened effective operation of government. Nixon aide John Ehrlichman outlined this argument in a briefing paper for the president. "We must classify some documents," Ehrlichman said, "*not* to deny information to those who have a need or right to know, but to permit government business to be successfully conducted, e.g., negotiations at SALT, Berlin, textile trade quotas, etc." He added, "Advisors must be able to rely on the confidentiality of their advice for a free flow of ideas." Ehrlichman concluded, "The classification law either applies or it doesn't. This obtains whether the subject is Vietnam, Iceland or atomic bombs. The law is the law whether the documents were prepared by people working for President Eisenhower or President Kennedy or President Johnson."[2]

The Supreme Court didn't see matters the Nixon administration's way, and a majority of the justices refused to countenance what they deemed pre-publication censorship. At this point, the administration's steps started to stray. Daniel Ellsberg, a former Pentagon analyst, had admitted by now to leaking the defense department's Vietnam study. The FBI had begun an investigation of the case, but the probe wasn't proceeding as quickly as the president desired. "I did not care about any reasons or excuses," Nixon wrote later. "I wanted someone to light a fire under the FBI in its investigation of Ellsberg, and to keep the departments and agencies in active pursuit of leakers." Nixon continued, "If the FBI was not going to pursue the case, then we would have to do it ourselves." In the antiwar atmosphere of the time, Ellsberg had become something of a celebrity. Nixon found this intolerable. "Ellsberg was having great success in the media with his efforts to justify unlawful dissent, and while I cared nothing for him personally, I felt that his views had to be discredited. I urged that we find out everything we could about his background, his motives, and his co-conspirators, if they existed."[3]

From this urging developed the White House "plumbers" unit, whose members broke into the office of Ellsberg's psychiatrist. From there to the Watergate complex in June 1972 was but another few steps. On the way, administration operatives tapped telephones, opened mail, and engaged in other illegal activities, all under the rubric of national security.

Like most presidents, Nixon had difficulty distinguishing between the interests of the country and the interests of his administration. It remains unclear which activities of the plumbers and other agents associated with the administration Nixon knew about in advance. But to the extent he did know about the Ellsberg-psychiatrist operation, or the wiretapping of individuals suspected of involvement in leaks, he doubtless had little trouble justifying these actions in terms of national security. As Ehrlichman had argued, the unauthorized disclosure of classified information, even if it didn't directly endanger the nation, would tend to stifle the honest exchange of opinions necessary to the formulation of policies needed to secure American interests.

The Watergate break-in, which amounted to plain political espionage, and the assorted dirty tricks the plumbers played, were harder to justify. The coverup that followed was harder still. Whether Nixon actually believed his own arguments is impossible to say—presidents in distress convince themselves of unlikely things—but he *claimed* national security as a justification for keeping federal investigators from looking too closely into the affairs of his administration and his re-election committee.

As happened continually during the postwar period, the ostensible demands of the Cold War were used to justify activities that bore only the most tenuous relation to national security. Under other circumstances, Nixon might have gotten away with his ploy. At the height of the anti-communist scare of the 1950s, his appeal to national security might have carried enough persuasive power for him to ignore his critics. (It would have helped his case, too, if he hadn't been such an unlikable individual.) But it was Nixon's bad timing that the Watergate plot thickened while he was in the middle of a process that took much of the bite out of the Cold War, thereby diminishing the credibility of the national-security excuse.

The road to Moscow, or Mao-maoing the flak catchers

In part, it was precisely Nixon's defanging of the Cold War that pushed him across the line into the illicit. Just as the *New York Times* commenced printing the Pentagon Papers, the president and his national security adviser, Henry Kissinger, were putting the finishing touches on the diplomatic spectacular that produced the most significant realignment in international affairs between the onset of the Cold War in the last half of the 1940s and the Cold War's end in the late 1980s. In July 1971, the Nixon administration amazed the world by announcing that Kissinger had just been to Beijing, where he had met with leaders of the People's Republic of China and arranged for a visit by the president to China early the following year.

Nixon's opening to China had two principal objectives, one immediate and the other longer-term. The immediate objective was to facilitate the American withdrawal from Vietnam in a manner that wouldn't precipitate the collapse of the government of South Vietnam. In 1954, at the climax of the French phase of the Indochina war, the Soviet and Chinese sponsors of the Vietnamese communists had prevailed upon Ho Chi Minh and his associates to accept less than they thought they deserved after their bitter struggle for independence. "We could have gained more," Vo Nguyen Giap, the commander of the Viet Minh army, asserted later. Pham Van Dong, the chief Viet Minh negotiator at the Geneva conference, declared flatly, "We were betrayed."[4]

Nixon was aiming for a second betrayal, if such was the right word. In particular, he hoped for efforts by the Chinese and the Soviets to persuade the North Vietnamese to settle for an incomplete victory, at least for the time being. The major communist powers had less pull with the Vietnamese communists in the early 1970s than they had had two decades earlier, since now the Vietnamese controlled a state of their own and the appurtenances thereof. Further, the rift that had developed between China and the Soviet Union rendered a united front toward Vietnam less likely than in the comradely first days of the Beijing-Moscow axis. Yet it seemed probable that should the Chinese and the Soviets withdraw their support for Vietnamese unification, from either common or competing reasons, Hanoi would have considerable difficulty achieving that long-sought goal.

Naturally, Nixon had to offer something to the Chinese and the Soviets in return for cooperation on Vietnam. To the Chinese, he offered improved relations, which would open the way for an increase in trade between China and the West (including that honorary Western country, Japan) and would otherwise strengthen China in its contest with the Soviet Union. To the Soviets, he offered arms control and an overall reduction in tension, both of which would ease the strain on the Soviet economy. To each, he offered the prospect of diplomatic and perhaps even military support against the other in the event the Sino-Soviet rivalry took a violent course. Implicit in this offer was the threat of opposition in the same event.

The second, longer-term objective of Nixon's opening to China was to rationalize American foreign policy by fitting Beijing into the framework of American diplomacy. During the early 1950s, when Soviet and Chinese international aims had run nearly parallel, a unified anti-communist American policy hadn't required gross distortions of reality. By the 1960s, however, the split between the Soviet Union and China had rendered simple anti-communism anachronistic—and not merely anachronistic, but counterproductive, since American hostility tended to keep Moscow and Beijing closer together than they otherwise would have been. In addition, where in the early 1950s the United States had commanded the strategic, economic, and psychological resources to at least contemplate garrisoning the entire non-communist world against communist expansion, by the end of the 1960s it had become apparent that such an objective was quixotic.

Nixon conceded this fact in a 1969 speech at Guam. Although here he spoke in the context of Vietnam and Asia, he later generalized the idea into what became known as the Nixon Doctrine. The United States would continue to protect its allies against the threat of nuclear attack by a major power, Nixon said, but it would no longer shoulder the burden of maintaining internal security for its friends, or of guaranteeing them against local or regional conflict. "The United States is going to encourage, and has a right to expect, that this problem will be increasingly handled by, and the responsibility for it taken by, the Asian nations themselves."[5]

Although Nixon declined to advertise the fact, this statement reflected, at least potentially, a significant turning point in the Cold War. To the extent that the statement would be implemented, it amounted to nothing less than a repudiation of the Truman Doctrine and of the notion that a victory anywhere for communism jeopardized American national security. Truman in 1947 had pledged American aid to non-communist regimes threatened "by armed minorities or by outside pressures." Nixon in 1969 was saying that unless the threat involved nuclear weapons—and it almost never did—those regimes had better look after themselves.

Even more fundamentally, what the Nixon Doctrine and the opening to China, and the subsequent policy of detente toward the Soviet Union, represented was an intention to de-ideologize American foreign policy. Until this time, American Cold War policy had embodied both ideology and geopolitics. It still would, to some extent. But by cozying up to the Chinese, Nixon was arguing that not all communists were bad, and not all had to be America's enemies. Nixon was trying to free the United States from the ideological straightjacket that had shaped its foreign relations for almost a quarter-century. Vietnam had demonstrated America's inability to guarantee the global status quo by main force. Now Nixon intended to substitute subtlety—in particular, the careful balancing of China against the Soviet Union—for force. If it succeeded, the new approach would allow the United States to maintain the international order and stability that had constituted chief objectives of American foreign policy since 1945.

News of the impending rapprochement between the United States and China stunned the world. The Soviet Union took a full week to formulate a response, and then the Kremlin's top America-watcher, Georgy Arbatov, warned that while an improvement in Sino-American relations might have peaceful motives, which the Soviet Union of course would applaud, it might also reflect hostility toward the homeland of socialism. The Soviet Union would examine future developments with close attention. "This is a matter of grave consequence for the Soviet people, for world socialism, for the entire international situation, for world peace," Arbatov said. Japan, unconsulted and uninformed of the American demarche, was flabbergasted by what the Japanese called the "Nixon shokku," although Japanese leaders politely kept most of their annoyance private. Eventually, Tokyo played a China card of its own, with Japanese prime minister Kakuei Tanaka traveling to Beijing, where he and Chinese premier Zhou Enlai agreed to terminate the state of war that had formally existed between the two countries since the 1930s, and to open diplomatic relations.[6]

In the United States, the reaction to Nixon's show-stopper was mixed. Many Democrats and liberals, including Democratic presidential hopeful George McGovern, found themselves having to applaud Nixon for doing what they would have preferred to criticize him for not doing. William Fulbright liked the results of the Nixon-Kissinger initiative, but predictably thought that it would have been "useful and appropriate" for the White House to have let the Senate foreign relations committee in on the administration's secret. Former China hands John Stuart Service and John

Paton Davies told Fulbright's committee that a move toward normalization with Beijing was long overdue. George Ball, Lyndon Johnson's undersecretary of state, agreed that normalization with China was tardy, but feared that in opening doors to China the Nixon administration would close doors to Japan. "Though we should develop relations with China within the narrow limits possible," Ball said, "our prime Asian objective should be to strengthen ties with a Japan not yet solidly anchored to the West." (To anchor Japan more solidly, the Nixon administration consented to return Okinawa and the rest of the Ryukyu Islands to Japan.) Dean Acheson conceded that a policy of nonrecognition of China was "outworn," yet the former secretary of state (and one of the fathers of nonrecognition) wasn't sure that the president's "rather precipitate and dramatic" approach was the right way to repair it. Acheson added, "Mr. Nixon is on a ledge pretty narrow for safety, his and ours." A writer for the Nixon-phobic *Nation* commended the president's "breakaway from two decades of unreality in regard to China," but cautioned that rapprochement with China might lead to a deterioration of relations with the Soviet Union. "One result of this could be a drastic heating up of the Cold War. Could it be possible that certain minds in the Administration, puzzling over how to win an election, would regard this as less than an unmitigated evil?"[7]

The liberals weren't the only ones with the 1972 election on their minds. Republican governor William Milliken of Michigan, worried that the Democrats would be able to use the uncertain state of the nation's economy against the GOP in 1972, welcomed Nixon's subject-changing visit to China as "a major coup." Other Republicans and conservatives were less impressed. Barry Goldwater, while refraining from criticism of his party's leader, displayed a marked lack of enthusiasm for the new policy. Congressman John Schmitz, who held Nixon's old seat in the House of Representatives, announced that he was "breaking relations" with the White House over the president's decision to improve relations with China. New York senator James Buckley decried the president's China announcements, while Buckley's brother William headed a group of eleven prominent conservatives, including *National Review* colleagues James Burnham and William Rusher, who declared that they had decided to "suspend" their support of the Nixon administration.[8]

Mixed reaction or not, Nixon plunged ahead. Lenin had once said that the road between Moscow and the West ran through Beijing. Nixon certainly found this to be true, for without his opening to China, the Soviet Union would have shown considerably less interest in bettering relations with the United States. Not that a reduction of tensions between the nuclear superpowers didn't make sense to the Soviets on its own terms. After a decade of devoting a major portion of their national resources to catching up to the United States in the arms race, the Russians were ready for a respite. Further, they had no greater wish to incinerate the world, by accident or design, than the Americans did. (In at least one respect they had less, since as official atheists they couldn't look forward to life after nuclear death.) But what put a particular edge on the Soviets' desire for detente was their fear that the Americans and the Chinese were ganging up on them. Preventing the formation of a solid Washington-Beijing alliance became, almost at once, a principal goal of Soviet diplomacy. Not even the intensification of American bombing of North Vietnam, or the mining of Haiphong harbor, deflected the Kremlin from this objective. Despite these provocations, Moscow refused to cancel a Soviet-American summit meeting scheduled for May 1972.

The atmosphere of the Moscow summit reflected the bland public personalities of the two summiteers, Nixon and Soviet general secretary Leonid Brezhnev. But, unusually for summits, the substance of this one mattered more than the atmospherics. Nixon and Brezhnev agreed to twelve "basic principles" that should guide Soviet-American relations. Though a few of the principles seemed added to round out the dozen, some involved issues of real importance. The first amounted to an ideological non-aggression pact. "Differences in ideology," it declared, "and in the social systems of the USA and the USSR are not obstacles to the bilateral development of normal relations based on the principles of sovereignty, equality, non-interference in internal affairs and mutual advantage." The second principle came close to being a military non-aggression pact: "The prerequisites for maintaining and strengthening peaceful relations between the USA and the USSR are the recognition of the security interests of the Parties based on the principle of equality and the renunciation of the use or threat of force." The third principle extended the first and second by asserting a "special responsibility" on the part of the United States and the Soviet Union to "do everything in their power so that conflicts or situations will not arise which would serve to increase international tensions." Waging wars by proxy, exporting revolution, and troublemaking by analogous means were now out of bounds.[9]

More specific and concrete than the declaration of principles were two arms control pacts Nixon and Brezhnev signed at the Moscow summit. The weightier of the twin SALT I accords banned comprehensive anti-missile defenses, allowing each side one installation near the national capital and one near a missile field. Because the erection of defensive systems would have spurred the deployment of additional offensive weapons to overwhelm the defenses, the ABM treaty implicitly put a check on the arms race in offensive missiles as well. The other SALT I agreement made this check explicit, capping offensive missiles and other delivery systems for five years, pending completion of a second round of talks.

Restructuring the Cold War

Together the basic principles and the SALT I accords formed the foundation for what might have been a new era in Soviet-American relations.

From the American viewpoint, the characteristic feature of the prospective new era was a willingness to deal with the Soviet Union as an ordinary great power. Soviet interests might differ from American, creating a certain degree of friction. And because the Soviet Union possessed the capability to annihilate the United States, American leaders had to treat Moscow with care and respect. But American leaders were backing away from the operating premise that Moscow was actively seeking world conquest, and they were content to handle Soviet-American affairs on a live-and-let-live basis. Though rivalry between the two great powers would continue, such rivalry would be less ideological than geopolitical, less a matter of broad existential purpose than of pragmatically specific give and take.

The new cooperative tendency had surfaced even prior to the Moscow summit. During the late summer of 1970, American intelligence sources detected unusual activity near the Cuban port of Cienfuegos. Construction crews had built a barracks and cleared ground for a soccer field. The latter bit of evidence, Kissinger later claimed, tipped him off at once. "As an old soccer fan I knew Cubans played no soccer," he wrote. "In my eyes this stamped it indelibly as a Russian base." (Actually, Cubans did—and do—play soccer, though not as enthusiastically or as well as they play baseball.) In addition, American planes had recently noted the arrival at Cienfuegos of a Soviet submarine tender and the rest of a naval group designed to service submarines. The CIA concluded that the Russians were developing a naval base in Cuba. "Soviet naval units," the agency asserted, "including nuclear powered submarines, may soon be operating regularly out of the Cuban port of Cienfuegos."[10]

Since submarines stationed at Cienfuegos might carry offensive nuclear weapons, this development appeared to portend a violation of at least the spirit of the agreement that had ended the 1962 missile crisis: no Soviet missiles in Cuba, and no American invasion of Cuba, with a collateral pullout of American missiles from Turkey. (In fact, that agreement had never been formalized, though each side had declined to test it thus far.) Had Nixon and Kissinger desired to make a major issue out of the Cienfuegos case, they might have gone public, as Kennedy had done with the missiles in 1962. Instead, they chose to handle it quietly. When reports of the base construction began to leak, Kissinger dodged questions. "We are watching the events in Cuba," he commented noncommittally. "We are not at this moment in a position to say exactly what they mean. We will continue to observe them and at the right moment we will take the action that seems indicated."[11]

Shortly thereafter, Kissinger met with Soviet ambassador Anatoly Dobrynin, and urged the Russian envoy not to misinterpret the administration's low-key public statement. Kissinger recalled the meeting in his memoirs: "I wanted him to understand that this was said only to give his government a graceful opportunity to withdraw without a public confrontation. We considered the construction at Cienfuegos unmistakably a sub-

marine base. Moscow should be under no illusion; we would view continued construction with the 'utmost gravity'; the base could not remain. We
would not shrink from other measures including public steps if forced into
it; if the ships—especially the tender—left Cienfuegos, we would consider
it a training exercise."[12]

While allowing the Kremlin the opportunity to think the matter over,
Nixon flew off to Europe on a trip previously planned. At the same time,
the president and top administration officials studiously ignored a flutter
of news stories about the base, and expressions of concern by conservatives who worried at this latest manifestation of a communist design for
world conquest.

When Nixon returned from Europe at the beginning of October, the
Soviets indicated a wish to liquidate the problem inconspicuously. Ambassador Dobrynin delivered a note restating Moscow's commitment to the
1962 understanding. "The Soviet side has not done and is not doing in
Cuba now—that includes the area of the Cienfuegos port—anything of
the kind that would contradict that mentioned understanding." The note
described certain American provocations, suggesting these might justify
Kremlin reconsideration of the 1962 deal, but then it returned to the point.
"In any case, we would like to reaffirm once more that the Soviet side
strictly adheres to its part of the understanding on the Cuban question and
will continue to adhere to it in the future on the understanding that the
American side, as President Nixon has reaffirmed, will also strictly observe its part of the understanding." (A few months earlier, the Soviets had
asked Kissinger whether the United States government considered the
1962 agreement to be still in force. Kissinger had checked with the president, and said it did.) Upon receiving additional assurances that the
United States would abide by its part of the 1962 bargain, Moscow
dropped whatever plans it had for a submarine base in Cuba. The Kremlin
procrastinated somewhat in pulling the tender and the other support ships
home from Cuba, but eventually the affair discreetly faded away.[13]

Nixon and Kissinger had a number of reasons for handling the Cienfuegos affair as they did. In the first place, the United States no longer
enjoyed the nuclear superiority it had in 1962. The 1970 business never
approached the shooting stage, but the president and his national security
adviser had to recognize that if it did, Moscow could shoot back with
considerably greater effect than it could have eight years earlier. As a
consequence, they understood their lesser capacity, compared with that of
Kennedy, to push the Russians publicly into a corner and make Moscow
comply with America's wishes. Second, only a few months before, Nixon
had sparked outrage all across America with the invasion of Cambodia.
While Kennedy in 1962 could count on the vast majority of Americans
rallying around the flag—and the president—in time of crisis, Nixon
couldn't, at least not with any great confidence. Indeed, skeptics such as
William Fulbright suggested that the Cienfuegos flap was a put-up job to

guarantee passage of the Pentagon's current budget request. Third, at just
the time of the Cienfuegos incident, a crisis developed in the Middle East.
Syria invaded Jordan, while the Jordanians fought the Palestine Liberation
Organization. With Moscow backing Syria, and Washington supporting
Jordan, the Nixon administration had to ponder the possibility of serious
trouble with the Soviets in the Levant. "They're testing us," Nixon told
Kissinger. Under the circumstances, an unnecessary fight in the Carib-
bean lacked appeal.[14]

Most important, a public challenge to the Kremlin might have loosed
the dogs of the Cold War in America, at a moment when Nixon and
Kissinger were trying to put them on a chain. The opening to China—the
fulcrum on which detente would rest—remained several months away,
and therefore subject to preemptive attack. For Nixon to pull off his Sino-
surprise required catching the hard-liners off guard, since they would
certainly cry Munich and Yalta and other nasty epithets if they thought
such action would block the undoing of twenty years of American policy
toward China. Once he got Beijing's agreement to move toward normalization, Nixon could present conservatives with a fait accompli. But until
then, he had to do whatever he could to keep the communist issue quiet.

On matters beyond Cienfuegos as well, Nixon and Kissinger attempted
to promote the idea that relations between the United States and the
Soviet Union need not be irredeemably antagonistic. In the run-up to the
Moscow summit of 1972, the administration began to relax strictures on
American exports to the Soviet Union, leading to, among other transac-
tions, a grain sale worth $1 billion. The success of the summit contributed
to further indications of economic cooperation: Nixon's pledge to allow the
American Export-Import Bank to extend credits to the Soviet Union, the
administration's promise to seek most-favored-nation status for the Soviet
Union, the opening of additional American ports to Soviet ships, and more
grain sales. In return, Moscow agreed to pay $700 million in Lend-Lease
debts from World War II, to open Soviet ports previously off-limits to
American shipping, and to undertake measures designed to encourage
development of Soviet markets for American products.

The economic aspects of detente weren't unimportant, certainly not from
the perspective of American grain farmers, ship-owners, and others who
hoped to do more business with the Eastern bloc. But to the way of thinking
of the White House, economics mattered chiefly in a larger context. Kissinger described the effort to increase trade with the Soviet Union as "an act of
policy, not primarily of commercial opportunity." Nixon delineated his
administration's attitude in a message to Congress. Speaking of commercial
ties between the superpowers, Nixon said, "They establish an interdepen-
dence between our economies which provides a continued incentive to
maintain a constructive relationship." Expanding his remarks to deal with
other forms of collaboration—political, technical, and cultural—the president stated that individual projects, taken separately, were "not crucial to

our relationship." But taken together they would "reinforce the trend toward more constructive political relations." The end result, he said, would be to "redirect the momentum of the past and chart a new direction in our relations with the Soviet Union, creating in the process a vested interest in restraint and in the preservation of peace."[15]

America's allies gave the Nixon administration a strong nudge in the direction it was going on its own. After the 1950s, the Cold War had never sold as well in the western half of the European continent as it had on the western side of the Atlantic, partly because the Europeans realized they might find themselves incinerated over some superpower dispute in Cuba or Africa or Southeast Asia. Nor did they like the second-class treatment they inevitably received in a world dominated by two superpowers. French president Charles de Gaulle particularly chafed at American hegemony in the West, and, calling for the creation of a single Europe "from the Atlantic to the Urals," in 1966 he pulled France out of NATO's unified command and ejected foreign troops, including American, from French soil. Lyndon Johnson recognized the futility of opposition. "When a man asks you to leave his house, you don't argue," Johnson told Robert McNamara. "You get your hat and go."[16]

The Germans also began moving out from America's shadow, aiming for a position more equidistant between Washington and Moscow. By the 1960s, West Germany's economic muscle was straining the Atlantic alliance, as German exports generated a trade surplus that threatened the stability of the dollar. In addition, the German revival sparked calls in Congress for the Germans to carry more of the burden of their defense, and for the United States, as a result, to carry less. Neither Kennedy nor Johnson wished to see the retrenchers gain momentum, since an American withdrawal would force the Germans to make their own arrangements for defense. American leaders liked this idea about as little as the Soviets did. First Kennedy and then Johnson postponed the day of reckoning through a series of "offset" deals, by which Bonn compensated Washington for stationing American troops in West Germany. This arrangement simultaneously neutralized the retrenchers and eased America's current account deficit. The accession of Willy Brandt in 1969 added another dimension to the drive for greater independence in German foreign policy. Brandt commenced unchaperoned negotiations with Moscow, leading to a 1970 non-aggression pact between the Federal Republic and the Soviet Union. The Nixon administration distrusted Brandt's policy. Kissinger, a refugee from Hitler's Reich, worried about German recidivism. "It seemed to me that Brandt's new *Ostpolitik*," Kissinger wrote, "which looked to many like a progressive policy of quest for detente, could in less scrupulous hands turn into a new form of classic German nationalism." But at a time when the United States itself was moving to diminish tensions with the Soviet Union, Washington couldn't much complain.[17]

Kissinger's remark indicated—accurately—that while the Nixon administration liked detente for the United States, it had reservations about detente for the allies. Phrased otherwise, though Nixon and Kissinger desired a relaxation at the top of the international pecking order, they preferred to keep the lesser chickens firmly in line. From the point of view of the United States, the Cold War entailed danger and expense, but it had the undeniable virtue of making the planet a reasonably orderly place. Ideology aside, the Cold War delineated, as clearly as one could expect in a messy world, the boundaries between the spheres of the superpowers. Within a remarkably short time, it evolved rules of conduct that guided relations between the only two countries that possessed the power to destroy each other and the rest of humanity. The rarity of direct confrontations between Washington and Moscow demonstrated the success of the Cold War system.

Of course, no system could survive unchanged indefinitely. The insight of Nixon and Kissinger lay in recognizing that the system needed modification, in particular to accommodate China. Actually, "insight" may be too generous a label: to most of the rest of the world, the need to accommodate China had been blindingly obvious for nearly a generation. In any case, accommodating China required softening the ideological constraints that had bound American policy since 1945.

But if Nixon and Kissinger were willing to retreat on matters of ideology, they were loath to surrender the orderliness of the Cold War framework. In his days in academia, Kissinger had grown enamored of Metternich and especially of the Austrian statesman's work at the Congress of Vienna, which seemed to Kissinger the model for structuring international relations. The less-intellectual Nixon apparently had no similar hero, but the preference of the two men for stability and order was unmistakable. A major purpose of the decision to regularize relations with China was to convert Beijing from being an anomaly in American foreign relations to being an ordinary part of the international landscape. To be sure, Beijing didn't fit into Yalta's bipolar scheme, but this fact had been evident, if publicly unrecognized by Washington, for a decade. The failure to recognize it had produced the Vietnam fiasco, which threatened to disorder American foreign relations utterly. By opening to China, Nixon and Kissinger would salvage what they could of the pre-existing framework of world affairs. Two poles might no longer be possible, but two and a half—China obviously being no match globally for the United States or the Soviet Union—might.

Even so, Washington had no desire to see the emergence of more poles than necessary. An independent Germany, unmoored from the West, could cause only problems. An independent-minded France already had. An unattached Japan would unsettle Asia. The Nixon administration's favorite euphemism for detente was "structure of peace." The emphasis was on structure, without which, in Washington's view, there would be no peace. The essential thrust of the third of the Moscow principles of detente—the one asserting the superpowers' "special responsibility" to "do everything in their power so that conflicts or situations will not arise which would serve to increase international tensions"—was that the United States and the Soviet Union would avoid shaking more than necessary the structure that had kept peace between the superpowers for a quarter-century.

Euphemisms notwithstanding, detente was a way not so much of structuring peace as of restructuring the Cold War. Nixon and Kissinger, realists to the point of cynicism, understood that peace was a relative matter. Peace was the avoidance of major conflict. Minor conflicts there would always be. Kissinger once put the purpose of detente simply: "We are in favor of detente because we want to limit the risks of major nuclear conflict." But the Cold War had had the same objective—which was why it was cold. Detente didn't end the competition that formed the basis for the Cold War. It simply restated the terms of the competition. Geopolitics, not ideology, henceforth would guide American policy. The competition wouldn't end until one side or the other thought the game was no longer worth the candle.[18]

It was understandable that Nixon and Kissinger should speak in terms of peace rather than ongoing Cold War. After a decade of Vietnam, American voters wanted to hear about peace, not more war. But by fuzzing up the issue of what they were doing, by overselling detente, the Nixon administration created expectations detente would have a difficult job fulfilling. As the reaction to his China policies indicated, conservatives eyed detente with skepticism. It would have to deliver a lot to assuage the conservatives' doubts. In light of liberals' chronic vulnerability on Cold War issues, the conservatives held the key to detente's acceptance in the United States. If they rejected it, a return to the old Cold War, or something very much like it, would be only a matter of time.

Side effects

Another weakness of detente was that it tended to treat countries other than the most powerful as minor pieces on a global chessboard. Such treatment, of course, was the idea of the "structure of peace," which wouldn't have been much of a structure if all countries could act independently. This weakness would manifest itself in the long term when certain minor pieces acquired the ability to act independently, and threatened to overturn the board. In the short term, it manifested itself in America's unconscionable treatment of some of those pieces.

The short-term weakness showed up during Nixon's third year in office. In the spring of 1971, Bengali separatists in the eastern half of Pakistan renewed earlier agitation for the establishment of an independent Bangladesh. The Pakistani government responded with extraordinarily repressive measures, sending the army into Dacca and killing tens of thousands, perhaps 100,000, civilians. The fighting spread across East Pakistan during the following weeks. As many as one million persons died, and more than three million fled over the border into India's West Bengal. This flood of refugees greatly alarmed India, already stretched trying to deal with West Bengal's chronic problems of poverty and over-population. When the Pakistanis, upset at India's unwillingness or inability to prevent the Bengali rebels from operating out of bases on Indian soil, foolishly invaded India, New Delhi took the opportunity to march on Dacca, routing the Pakistanis and springing Bangladesh from Islamabad's grip.

This tragic sequence of events caught the Nixon administration in the middle of arranging the opening to China. The administration's primary contact in making the arrangements was Pakistani president Agha Yahya Khan. Yahya had secretly relayed to China Washington's desire for an improvement in relations, and when Kissinger slipped silently into Beijing in July 1971, he did so on board a Pakistan Airlines jet specially sent by Yahya. Predictably, Nixon and Kissinger didn't wish to upset Yahya, who could have blown the whole business and wrecked the administration's plans for reshaping the international structure of power. As a consequence, Washington acquiesced in the Pakistani government's shockingly brutal treatment of its own citizens, inflicted with weapons furnished and paid for by the United States. The reason for the acquiescence became apparent after Kissinger's trip to China, but by then the damage—to the hundreds of thousands of Bengali dead, to the millions homeless, and to America's good name—had been done. Previously, when the United States had consorted with authoritarian regimes, the consorting at least had had the objective of preventing the spread of totalitarian communism. Now it resulted simply from a desire to play one communist regime against another. A more machiavellian policy was hard to imagine.

Critics of the Nixon administration certainly thought so. In the Senate, the leaders of the Democratic party pounded the administration. Walter Mondale introduced a resolution to bar deliveries of American weapons to Pakistan, saying, "There is something very wrong when guns, tanks and planes supplied by the United States are used against the very people they are supposed to protect. There is something very wrong with a military aid policy that lends itself to this travesty." Frank Church denounced the Nixon administration for "aiding and abetting a terrible massacre on the part of the West Pakistan military regime." William Proxmire lambasted the administration for complicity in "genocide." Edward Kennedy returned from a visit to the war zone with graphic descriptions of the horrors visited upon the living, not to mention the murders of those now dead. "The situation in East Bengal should particularly distress Americans, since it is our military hardware—our guns and tanks and aircraft delivered over a decade—which are contributing substantially to the suffering. And even more shocking is the fact that these military supplies continue to flow—apparently under instructions from the highest officials of our land. Pakistani ships loaded with U.S. military supplies continue to leave American harbors bound for West Pakistan." Kennedy demanded that the administration change course at once. "We must end immediately all further U.S. arms shipments to West Pakistan. We must end all other economic support of a regime that continues to violate the most basic principles of humanity. We must demonstrate to the generals of West Pakistan and to the peoples of the world that the United States has a deep and abiding revulsion of the monumental slaughter that has ravaged East Pakistan."[19]

Major organs of the media joined the criticism. The *Washington Post*, asserting that "in Pakistan the world is witnessing a holocaust unmatched since Hitler," called for a halt to American military aid to the Pakistani government. The *New York Times* conceded that humoring Yahya might have made sense while the administration was planning the Kissinger trip to Beijing. "But now that the door to China has been opened, it is impossible to excuse or explain Washington's continuing supply of arms to Islamabad and its persisting ambiguity in the face of a deepening tragedy."[20]

Such attacks from the left on the Nixon-Kissinger foreign policy weren't particularly significant at the time. For vehemence and volume, they hardly registered next to the continuing liberal criticism of the administration's handling of the war in Vietnam. But the alienation of the liberals would matter later in the decade. The liberals were the natural constituency for detente, and, on the whole, they approved of the administration's non-ideological approach to relations with the communist powers. Yet the evident cynicism of detente's implementation at the hands of Nixon and Kissinger—combined with the liberals' thorough distaste for Nixon as an individual—left them less than enthusiastic about the president's policies. When the conservatives struck back against detente, the liberals would scatter quickly.

Post-traumatic stress syndrome

Six months after the India-Pakistan-Bangladesh war, the administration's bag men did their Watergate business, and a short while later the obstruction of justice began. The scandal that eventually came to light had a two-stage effect on American foreign policy, especially that part relating to the Cold War. In the first stage, it continued the de-escalation of the Cold War that detente had begun, although it did so with an emphasis the chief detentist, Nixon, hardly approved. In the second stage, it led to a re-escalation of the Cold War and a reassertion of the premises on which the Cold War rested.

The first stage showed most clearly with respect to Vietnam. Nixon's efforts to get the Soviet Union and China to persuade North Vietnam to accept a compromise peace proved considerably less efficacious than he had hoped. Moscow and Beijing did their part, if not with the conviction

Washington wished, but Hanoi resisted the pressure. The North Vietnamese complained loudly that unnamed but unmistakable socialist powers
were "bogging down on the dark, muddy road of compromise" and were
"departing from the great, all-conquering revolutionary idea of the age."
Evidence indicated that such powers had chosen "peaceful coexistence
over proletarian internationalism," and opted for "their own immediate
interests at the expense of the revolutionary movement." North Vietnam
wouldn't follow this misguided policy. "We Communists must persist in
revolution and should not compromise."[21]

When Soviet and Chinese pressure failed to produce a peace accord,
Nixon stepped up American pressure. At the end of 1972, he ordered the
most intensive bombing campaign of the war—in fact, the most intensive
bombing campaign in world history. The result was equivocal: the North
Vietnamese signed a peace agreement in January 1973, but in doing so
they gave up nothing of the substance of their long-standing goals. On the
crucial question of troops on the ground, the Americans withdrew while
North Vietnamese forces remained in place in South Vietnam. Nixon
pledged continued American backing for South Vietnamese president
Nguyen Van Thieu, but Thieu, who could read the declining support for
the war in the United States as well as the next person, wasn't counting
very heavily on it. To the end of the negotiations, American prisoners of
war held by the communists had served Hanoi as hostages. North Vietnam refused to repatriate the prisoners until the United States agreed to
leave Vietnam. Though Thieu had sufficient savvy not to say so, he
realized that the prisoners acted as hostages on his behalf as well, guaranteeing continued American interest in Vietnamese affairs. The Paris accord returned the prisoners, leaving Thieu as hostage-less as Hanoi. For
what little it was worth, he refused to sign the accord.

Events demonstrated the accuracy of Thieu's misgivings. After the
POWs returned home and the last American troops departed, the ranks of
those Americans concerned about the fate of South Vietnam dwindled
drastically. During 1973, Congress, showing the courage that comes from
careful monitoring of public-opinion polls, repealed the Gulf of Tonkin
Resolution and otherwise registered disapproval of fresh efforts to save
Saigon. A strong and determined president might—just might—have
overcome this disapproval, but by the end of 1973, Nixon was far from
strong, and his determination focused on staying one step ahead of Sheriff
Sam Ervin and the Watergate posse.

Nixon's downfall in August 1974 sealed Thieu's fate. After North Vietnam discovered—to no one's surprise—a pretext for overturning the
ceasefire, Congress refused Gerald Ford's request for emergency aid to
stem the communist advance. South Vietnamese resistance evaporated at a
speed that amazed even Hanoi. Saigon surrendered in April 1975.

The final defeat of anti-communist forces in Vietnam marked the low-
water point of American enthusiasm for the Cold War. In late 1973, Con-

gress had passed the War Powers Resolution, thereby limiting a president's ability to conduct military operations overseas without legislative consent. Whether such an act would have had a significant influence on American policy in Vietnam—its obvious, though ex-post-facto, target—was questionable. If Congress had voted for a declaration of war, as it probably would have if asked at the time of the Tonkin resolution, then the war would have become the legislature's as much as the president's. Under this dual proprietorship, disowning the war would have been even more difficult than it turned out to be. As for the future of the war-powers measure, presidents from Nixon onward consistently judged it unconstitutional: a usurpation by the legislature of the prerogatives of the commander-in-chief. Many knowledgeable observers deemed the act politically unenforceable, since the principal influence Congress can exercise regarding troop deployments is the threat of cutting off funds, and only a very bold or thoroughly alienated legislature will risk charges of leaving American boys defenseless in the face of hostile fire. Even so, the War Powers Resolution signaled a repudiation, if principally symbolic, of the kind of executive-dominated decision-making that had characterized American diplomacy during the Cold War.

Of greater substance was the legislature's refusal to get involved in another distant civil war, this one in Angola. After the outbreak of revolution in Portugal in April 1974, independence for Portugal's colonies in Africa was assured, and soon. A three-way contest for power in Angola quickly developed, but the fight only acquired strong international overtones in the first half of 1975, when Cuba began sending military advisers and then combat units to support the Marxist-oriented MPLA. South Africa joined the tussle at about the same time, invading in cooperation with the two other contending groups, UNITA and the FNLA.

Until nearly the end of 1975, the United States participated in the Angolan conflict primarily by means of covert aid supplied to the UNITA-FNLA coalition. By November of that year, the Ford administration had run through a few tens of millions of under-the-table dollars. But the well was going dry, and to keep the money flowing, the White House appealed to Congress. The administration had three reasons for wanting to prevent a victory by the side Moscow favored. The first was strategic. Detente or no detente, the United States had to worry about the establishment of a pro-Soviet regime in southern Africa. The second was diplomatic. For all Moscow's embrace of detente's principles, Kissinger understood that the Kremlin would probably test the boundaries of the permissible, pushing here and there to see if the Americans would push back. A firm posture in Angola, far from contradicting detente, would strengthen detente by letting the Russians know that the Americans were serious about preventing violations. The third reason was political. Those conservatives who had distrusted detente from the start were more convinced than ever by the Angola affair that detente was a dead, and deadly, end. If the Ford adminis-

tration intended to see detente survive, American officials needed to show
that it didn't entail handing all the unclaimed regions of the world over to
the reds.

Three reasons—these three, at any rate—weren't enough for Congress.
After Watergate, the dominant Democrats weren't going to let the Republicans get away with anything covert, and after Vietnam, they weren't
about to march off into any more jungles. Senator Birch Bayh set the tone
of the Democrats' argument. "Despite the tragic and bitter lessons we
should have learned from our intervention in civil war in Southeast Asia,"
Bayh declared, "the Ford administration persists in deepening our involvement in the civil strife in Angola." The Indiana lawmaker continued, "The
war in Angola presents the first test of American foreign policy after
Vietnam. Unfortunately, our performance to date indicates that we have
not learned from our mistakes. Rather than recognizing our limitations
and carefully analyzing our interests, we are again plunging into a conflict
in a far corner of the world as if we believed it was still our mission to serve
as policeman of the world." Thomas Eagleton of Missouri challenged both
the rationale for getting involved in Angola and the means by which the
president and his administration were effecting that involvement. "Is it
sound policy to become embroiled in an African war?" Eagleton asked. "Is
it wise to ally ourselves in this venture, even indirectly, with South Africa?
Are the deepwater ports of Angola and the shipping lanes off its coast
important enough to our national interests to warrant our involvement in a
war?" Eagleton thought not. Robert Byrd echoed Eagleton's concerns
about associating with South Africa. "Were the United States to become
jointly allied with South Africa in this conflict," Byrd predicted, "it could
spell disaster for our future relations with most of the states of black
Africa, as well as with the rest of the third world." Edward Kennedy
turned the old domino theory back on those who used it to support American intervention in foreign conflicts. "If ever there is a 'domino effect,'
Kennedy asserted, "it is how covert activities can easily fall into overt
involvement and long-term commitments." The Massachusetts senator
added, "The American people will not tolerate this. The United States
must avoid this in Angola."[22]

The mood in the House of Representatives was similar. Andrew Young
of Georgia decried administration policy in Angola, and stated, "To continue along the path of covert or overt intervention there will be a monstrous mistake for our national destiny and peoplehood." Henry Waxman
of California called the administration's support for UNITA and the
FNLA a "knee-jerk reaction" to Soviet support for the MPLA. "They
have become our allies only because they are fighting a group which is
backed by the Russians, and not because they promise the establishment
of a democratic and free country in that former colonialized nation." This
reasoning hadn't washed in Vietnam, and it wouldn't wash in Angola.
Christopher Dodd of Connecticut declared that American policy toward

Angola was "neither just nor reasonable." Dodd condemned Soviet meddling in Angolan affairs, but he didn't see that the Kremlin's crime required the United States to sin similarly. America mustn't get suckered into "playing the Soviet game" or "using identical tools in a show of force." Prestige counted, Dodd said, but not the reactive and reactionary pseudo-prestige of the Cold War. "It is high time that our foreign policy cease to be a creature of Pavlovian reflexes and for it to change in accordance with changing circumstances. It is vital to our image and our national well-being that we no longer fight side by side with the forces of oppression"—in this case, "the unabashedly racist South African regime"—"thus identifying ourselves with them in the eyes of the third world." Don Bonker of Washington, the sponsor of an amendment blocking aid to the Angolan factions, was pleased to read into the legislative record editorials from the *Washington Post* and the *New York Times* that supported his effort. Bonker quoted the *Post* as taking Dodd's line regarding Soviet intervention: "If the Russians are so perverse as to proceed, let them then harvest the scorn of much of the international community, the burden of carrying Angola if their client prevails, and the embarrassment of 'losing' Angola if their client's gratitude subsequently falters, as usually happens in these situations." The *Times*, with so many others, compared Angola to Vietnam: "In light of recent tragic experience, Americans are entitled to ask, 'Where will Washington's side of the escalation end?' "[23]

The administration's critics carried the day with ease. When the voting came, the opponents of the requested aid won by a 54 to 22 margin in the Senate, and by 323 to 99 in the House.

This reaction to the Vietnam debacle indicated a new twist in the American history of the Cold War. Previous setbacks had simply reinforced America's commitment to containment. The communist victory in China in 1949 had provoked immediate demands for renewed resistance to communist advances in Asia, leading within the year to American intervention in Korea's war. Castro's victory in Cuba in 1959, and his subsequent turn to communism, had triggered redoubled efforts against communism in the Western Hemisphere, leading to the Bay of Pigs operation and the showdown over the Cuban missiles. Efforts to contain communism in Latin America also included the 1965 American invasion of the Dominican Republic. By contrast to these reassertions of American anti-communism, the reaction to the fall of South Vietnam was chiefly a determination to avoid similar mistakes in the future, by avoiding the kind of commitments that had produced this one.

To some degree, the no-more-Vietnams attitude reflected the success of the detentists in de-ideologizing the Cold War. Having shared tea with the Chinese communists and bear-hugged the Russians, it was difficult for the Nixon-Kissinger-Ford crowd to generate concern lest some Marxists gain an edge over their rivals in a civil war in an obscure country in Africa. To some degree, the American aversion to future Vietnams reflected moral and psychological exhaustion. Not all the wrenching of the 1960s and early 1970s had resulted from Vietnam, but much had. And what hadn't, often seemed to have, such being the way of bad memories. Most Americans entertained no desire to repeat the ordeal. To some degree, the no-more-Vietnams sentiment reflected a salutary appreciation of the limits of America's ability to shape the world according to American designs. Americans had tried in Vietnam and failed. They saw no reason to court like failure elsewhere.

Not least, the lack of enthusiasm for the ambitious policies that had produced Vietnam, and that had marked America's approach to the Cold War from the start, reflected a widespread disillusionment regarding what the Watergate-inspired investigations were revealing about the manner in which the United States had fought communism during the previous twenty-five years. When Congress discovered that the FBI and the CIA had participated in Nixon's attempted coverup, congressional committees began asking what else those organizations had been doing. In the case of the FBI, the death of Hoover in May 1972 facilitated the excavating. (One wit thought Hoover's passing so significant as to label Watergate the "war of the FBI succession.") It was amid the revelations of wrongdoing in the Nixon White House that the nation learned of the FBI's campaign against the Socialist Workers and other radical and not-so-radical groups during the 1960s. This information set researchers, armed with a newly fortified version of the 1966 Freedom of Information Act, on the trail that led to the release of the bureau's massive file on Martin Luther King.[24]

The digging on the CIA turned up even more dirt. Senator Frank Church's congressional committee discovered that American agents had conspired to assassinate Patrice Lumumba and Fidel Castro; that the CIA had tested mind-altering drugs, including LSD, on unwitting subjects, apparently driving at least one subject to commit suicide; that the agency had regularly violated its charter in conducting intelligence operations in the United States; that it had had contravened American law in opening mail in the United States; and that it had engaged in sundry other activities it shouldn't have. The congressional investigations provoked a spate of books, several by changed-of-heart former insiders, recounting CIA involvement in the overthrow of Mossadeq in Iran, Arbenz in Guatemala, and Salvador Allende Gossens in Chile, among its more noteworthy king-unmaking efforts.

The public response to the revelations ranged from shock to what-do-you-expect, with Cold War loyalists warning against throwing babies out with bathwater. Opinionists on the left reveled in the muck. The *Progressive* wondered why everyone else had taken so long to catch on to what it had known and been telling its readers for decades. "The problem that confronts us," the magazine declared, "is not just that the CIA occasionally lapses into illegal activities, but that it is, inherently, an illegitimate enterprise. No institution cloaked in secrecy, immune to scrutiny, and empowered to spend vast sums of public money without accounting to the taxpayers or even to their elected representatives can constitute anything but a threat to democracy." The *Progressive* doubted that the current flurry of investigations would yield any serious reforms. "A few wrists will be slapped, a few bureaucrats will be replaced, a few pronouncements will be issued to assure us that the Republic is safe. It is not, and it will not be." Absent a radical reordering of American priorities—detente hardly qualified—the imperatives of the national security state would continue to claim precedence over those of democracy.[25]

The *Nation* saw matters in much the same light. James Higgins, reviewing *The CIA and the Cult of Intelligence* by Victor Marchetti and John Marks, described the CIA as an almost-wholly owned subsidiary of American corporatism. Higgins asserted, "This has to be clear: that the CIA has functioned, bankrolled by billions of public money to be sure, in behalf of private economic interests—of what used to be called simply big business." Higgins examined the CIA's antecedents and continued, "The Central Intelligence Agency was founded on cold-war premises, assigned to beat the bad guys by hook or crook, advised that the phrase 'national security' would be employed on the highest levels to justify any damn thing that went right or wrong—expected, in short, to serve as roving agents of the policy of corporate geographical expansionism which had its origins in the Westward-ho era of the 19th century, initiated almost as soon as the Civil War came to an end." Higgins held that the instance of Chile, which had hardly been under assault from Soviet communism, but whose government had evinced an indubitable antagonism toward foreign multinationals, demonstrated plainly that the argument from national security was simply a fig leaf for corporate privateering. This fact, he said, explained the extraordinary efforts the Ford administration had made to keep the Chile affair secret.

> No wonder the government went all-out to delete Chile references from the Marchetti-Marks book. No wonder Ford had to step (be pushed?) forward to repeat, as in an echo chamber, national security nonsense to support a case for CIA interference in Chile, once the deleted matter came to light in reports by [Seymour] Hersh, [Laurence] Stern and others. The point is that Chile, at one and the same time, implied the old national security argument to have been a historical lie—a big lie, to use the words once applied to Nazi deception of the Germans when big business in Germany sought domination of world resources and people—and also threatened to reveal that under cover of competitive coexistence with the giants of the Socialist camp, the CIA and its masters intended to continue playing dirty tricks wherever possible, their aims being necessarily less grand than the Nazis', and their techniques more sophisticated, but both aims and techniques comparable in design to what Hitler's backers had in mind.[26]

It wasn't especially noteworthy that journals like the *Progressive* and the *Nation* were saying such things about the American intelligence establishment and the Cold War philosophy on which it was based. This was old news. What *was* noteworthy was that lots of other journals and people were listening to them, and saying similar things themselves. The *New Republic*, closer to the middle of the American political spectrum, carried an article by Roger Morris agreeing that the CIA's alleged excesses weren't excesses at all, but the logical product of the Cold War system. "As Congress and the public recoil from many of the CIA's past actions, from murder to political mayhem, they are also seeing a faithful, albeit sordid reflection of American foreign policy over the last quarter century." Morris likewise agreed that current efforts to reform the system would probably fail, although he saw bureaucratic banality rather than corporate venality as the fix-foiling culprit. "The real stopper to genuine reform is not some stentorian guard of cold-war assassins, but rather what mires reform all over Washington—that legion of little people, whose cloaks are knit suits and daggers are government-purchase cafeteria butter knives, with suburban mortgages, children in college, and lives invested seemingly beyond return." Congress might, possibly, take advantage of the present fleeting opportunity to set America's intelligence system right. But it would have to show more courage than it had to date. "The investigation must dig deeper and wider than Congress has been willing to go thus far, and the senators and representatives must be prepared to confront not only generally condemned aberrations, but also equally repulsive products of business as usual."[27]

The editors of *Newsweek*, more centrist still, placed at the front of their magazine an essay by journalist A. J. Langguth with the straightforwardly emphatic title "Abolish the CIA!" Langguth didn't deny that the CIA provided certain necessary services to the United States. He went so far, in fact, as to concede that 80 percent of the agency's activities were worthwhile. But this 80 percent lay entirely in the area of intelligence research and analysis, and the vast majority of it drew on openly published newspapers, magazines, and the like. The people who did this work could just as easily do it above board in the State Department. As for certain other information sources, such as satellite photographs, Langguth contended that these didn't require secrecy. "I'd like to see us routinely make our prints available to the world," he said.

So what would he rule out? "The sport. Simply, the sport of grown men with the nature of an E. Howard Hunt. The sport of traveling under code names, bribing foreign officials, collecting the material that overflows the computer banks at Langley, Va." Langguth granted that forswearing such activities would come hard. "A generation warped by the cold war may find the idea of giving up the CIA's covert or 'black' operations almost suicidal." But only in this way would the American people succeed in calling to account those who directed American foreign policy. "When it is deemed vital to legitimate American interests that an Allende go in Chile, let the case be put to Congress. Then, send in the Marines or the airborne. . . . From here on, let the U.S. President who orders such enterprises be ready to sell them to Congress and to his constituents, or face an impeachment process, which, we have learned, is not so traumatic after all."[28]

Criticism of the CIA ranged well into conservative haunts. *Business Week* didn't buy the running-dogs-of-capitalism reasoning, but the magazine's reviewer gave remarkably sympathetic treatment to Philip Agee's *Inside the Company*—which was all the more remarkable considering that Agee fingered nearly everyone he had known or worked with in the CIA, placing the fingered in undeniable danger. The review described Agee's story as a "dispiriting account" that would force Americans to revise their thinking about what their country actually stood for, American rhetoric notwithstanding. "For many Americans who believe in Jeffersonian and Wilsonian democracy, it presents a documentary lesson in why CIA activities have made the U.S. so feared and disliked overseas by so many who should regard this country as their friend and hope."[29]

Even William Buckley's *National Review* added to the critical chorus. James Burnham, the original neoconservative, whose hegira from left to right in the 1940s inspired a rush that continued for a generation, declared himself convinced by the evidence against the CIA. "There's no doubt about it," Burnham confessed. "Hubris does come before a fall. I would have boasted myself one of the last persons on earth who could have been snowed by the CIA. And I've just learned that those rascals hornswoggled me as neatly as if I'd been a college president." After reading the Marchetti and Marks book, Burnham accepted much of the authors' indictment. The American intelligence system, he said, suffered from "acute elephantiasis." The CIA and its sister agencies—the National Security Agency, the Defense Intelligence Agency, the intelligence arms of the military services, to name a few of many—tended to pursue intelligence because it was there, not because someone had demonstrated that the intelligence had value. The spy agencies ordered the latest and most expensive gadgets for similar reasons.

Nor did covert operations escape the Everest imperative. "The fact that a covert action (dirty trick) section exists in CIA," Burnham wrote, "creates an inevitable wish to assign it missions even if it is not clear how those missions serve the national interest." What made matters worse was that covert operations subverted the intelligence mission of the CIA. "The co-presence of a covert action branch and an intelligence branch in the same organization, as is the case in CIA, inevitably means that intelligence analysis will be adapted to justify the covert action." Burnham cited the Bay of Pigs affair as a prime example. "Prior to the Bay of Pigs, the intelligence estimates kept insisting that the Cuban people would rise en masse against Castro if given a chance, not because this was true but because it was necessary to the rationale of the anti-Castro expedition." Under such conditions, wise decision-making was nearly impossible.

Like other critics, Burnham argued that the secrecy was greatly over-done. "There is no real reason why many of the covert actions should be covert, especially when, as is the case of most of them, everyone interested knows who is involved anyway." Again like other critics, Burnham doubted that the current investigations would produce meaningful reform. For reasons psychological and political, the investigators favored sensa-tional exposés over the harder but more prosaic work of finding real solu-tions to the problems of intelligence. "It is much more fun to talk about assassinations," he concluded.[30]

Carterizing the wound

As Americans learned what their government had been doing in the name of containing communism, their enthusiasm for the Cold War diminished further. The Church committee summary included excerpts from James Doolittle's 1954 report, the one that had asserted that the United States must "subvert, sabotage and destroy our enemies by more clever, more sophisticated and more effective methods than those used against us." Having believed for nearly three decades that the United States adhered to a higher standard in the Cold War than the communists did, most Ameri-cans found it distressing to discover that the same rules appeared to apply on both sides of the Iron Curtain. A few judged the United States the greater culprit, but this was principally because the Soviets at this junc-ture weren't admitting anything.

Amid the disenchantment, Americans elected Jimmy Carter, the Demo-crats' only winning presidential candidate in the period between the Americanization of the Vietnam War and the end of the Cold War. To a certain extent, Jimmy Carter was a self-limiting phenomenon. He ran as an outsider, a person not tainted by the corruption of incumbency. Once elected, he lost this advantage to the next anti-government candidate to come along. (Ronald Reagan would trot out anti-government themes for a second time in 1984, but the question that won him re-election—"Are you better off now than four years ago?"—showed his true incumbent colors.)

Whether Carter missed the point of detente as conceived by Nixon and Kissinger, or whether he understood and just didn't like their geopolitical version of the Cold War, he soon set about designing a policy that would have consigned the Cold War to the past. Early in his term, he outlined his foreign-policy objectives in a widely noted speech at Notre Dame Univer-sity. Carter contended that Americans—and Russians, for that matter—must recognize that their days of ordering other countries around were past. "We can no longer expect that the other 150 nations will follow the dictates of the powerful," he asserted. The war in Vietnam, which had produced "a profound moral crisis, sapping worldwide faith in our own policy and our system of life," should have convinced Americans of this fact. It should also have convinced Americans of the folly of focusing on communism and the Soviet Union to the exclusion of other forms and perpetrators of oppression. Carter believed it had. "We are now free of that inordinate fear of communism which once led us to embrace any dictator who joined us in that fear. I'm glad that that's being changed."

Carter thought this last theme worth elaborating. "For too many years," he said, "we've been willing to adopt the flawed and erroneous principles and tactics of our adversaries, sometimes abandoning our own values for theirs. We've fought fire with fire, never thinking that fire is better quenched with water. This approach failed, with Vietnam the best example of its intellectual and moral poverty. But through failure we have now found our way back to our own principles and values."

Carter stressed that the world of the 1970s wasn't the world of the 1940s. The superpower conflict wasn't the only conflict that mattered, and relations among the advanced states weren't the only relations worth bothering about. "We can no longer have a policy solely for the industrial nations as the foundation of global stability," the president declared. "We must respond to the new reality of a politically awakening world." He added, "We can no longer separate the traditional issues of war and peace from the new global questions of justice, equity and human rights."

A policy to promote justice, equity, and human rights couldn't be a policy designed according to the ideas and methods that had typified the Cold War. "We cannot make this kind of policy by manipulation. Our policy must be open; it must be candid; it must be one of constructive global involvement." Carter specified some cardinal features of such a policy. The first was a "commitment to human rights as a fundamental tenet of our foreign policy." The president didn't deny the limits of moral suasion, and he said he had no illusions that change would come easily or soon. But still he asserted, "It is a mistake to undervalue the power of words and of the ideas that words embody." He stated that those who placed military strength and economic prosperity ahead of human freedom had things just backward. "The great democracies are not free because we are strong and prosperous," he said. "We are strong and influential and prosperous because we are free."

A second fundamental feature of his foreign policy, Carter explained, would be an active effort to halt and then reverse—not simply slow—the strategic arms race. "This race is not only dangerous, it's morally deplorable. We must put an end to it." Getting to details, he declared, "We desire a freeze on further modernization and production of weapons and a continuing, substantial reduction of strategic nuclear weapons." In addition, he called for a comprehensive ban on nuclear testing, and for prohibitions against chemical warfare and the deployment of anti-satellite weapons.

As a third point, Carter emphasized the need to deal with the growing disparity in living standards between the prosperous nations and the poor, to shift attention and resources away from the East-West issues that had dominated the Cold War and toward the North-South questions that would shape the future. "More than 100 years ago, Abraham Lincoln said that our country could not exist half slave and half free. We know a peaceful world cannot long exist one-third rich and two-thirds hungry." Through technical cooperation, trade liberalization, education, and more-effective aid programs, the developed countries should assist the less-developed in closing the affluence gap.

Carter acknowledged that the United States might at times have to resort to military force to secure American interests around the world. But his emphasis was definitely in the direction of noncoercive instruments of persuasion. "Our policy is rooted in our moral values," he declared. "Our policy is designed to serve mankind. And it is a policy that I hope will make you proud to be Americans."[31]

What Carter described was a foreign policy for a post–Cold War world, a policy with an agenda considerably broader than that which had informed the Cold War. It placed positive goals such as justice and human rights ahead of the negative objective of containing communism and the Soviet Union. It embraced change, instead of opposing or merely tolerating change. It argued for the force of American example, rather than the force of American arms, as the principal guarantor of American interests. It called on Americans to live up to their own moral values, not to live down to those of others.

To a certain degree and for a certain time, Carter's vision of a world beyond the Cold War echoed elsewhere in the American political system. Congress shared the president's desire to promote human rights. Amid the post-Watergate revelations of American manipulation of foreign governments—and with the sour taste of the Pakistan-Bangladesh experience still in many mouths—the legislature approved a resolution declaring that expansion of the area of regard for human rights should be "a principal goal of the foreign policy of the United States." To this end, Congress directed the secretary of state to evaluate the human rights records of all governments receiving American aid, and specified that gross and consistent violators be purged from the roll of recipients.[32]

Carter went beyond this directive. He created a special State Department post for monitoring human rights, and filled it with Patricia Derian, an ardent civil-rights advocate from Mississippi. He corresponded with Soviet dissident Andrei Sakharov, despite the annoyance the correspondence caused the Kremlin. "I am always glad to hear from you," Carter wrote Sakharov. "You may rest assured that the American people and our government will continue our firm commitment to promote respect for human rights not only in our country but abroad." He signed United Nations covenants on civil and political rights, and on economic, social, and cultural rights, and lauded the measures as "tangible steps toward the realization both of peace among nations and the preservation of human rights for individual men and women throughout the world." He ordered cutbacks or cancellation of American aid to repressive regimes in Argentina, Chile, and Nicaragua. He implemented trade sanctions against the Ian Smith government of Rhodesia and the Idi Amin government of Uganda. He banned the export of weapons and police equipment to South Africa. He directed American diplomats to advise the governments of other countries—Iran, the Philippines, South Korea—to shape up. He ensured a high American profile at conferences summoned in pursuit of the Helsinki human-rights accords of 1975.[33]

Carter likewise applied himself to the subject of arms control. Building on the SALT I agreements of the Nixon administration, and on a subsequent interim pact approved by Ford and Brezhnev at Vladivostok in 1974, the Carter administration pressed for lower limits on strategic arms. The administration's chief negotiators, Paul Warnke and Secretary of State Cyrus Vance, proposed to cut superpower arms to levels considerably below those specified at Vladivostok. The Soviets initially refused, suspecting an American trick—Why the sudden change? The Soviets also objected to the fact that they would have to do the sharpest cutting, while the Americans would simply have to shelve future building plans. But eventually the two sides agreed to modest reductions in launchers (from 2,400 on each side to 2,250), and to ceilings on other parts of the nuclear arsenals. The resulting SALT II agreement was signed by Carter and Brezhnev in Vienna in 1979.

Carter demonstrated his desire to de-emphasize the East-West issues of the Cold War in favor of the North-South concerns of the Third World by, among other actions, appointing Andrew Young to be the American ambassador to the United Nations. Young had first gained widespread attention as a lieutenant to Martin Luther King in the civil rights movement. He brought to the United Nations post a sensitivity to matters of importance to Africans and Asians rarely apparent in previous administrations. Young had little patience for the Cold War argument that the Soviet threat in Africa necessitated American collaboration with the apartheid regimes in South Africa and Rhodesia, or required American support for the anti-Soviet and anti-Cuban rebels in Angola. He contended that such thinking produced precisely what it aimed to prevent: that by siding with despised reactionaries, the United States opened Africa to Soviet adventurism. Young also urged a more equitable distribution of the world's wealth as a means of alleviating the despair that engendered support for communism, and of ensuring greater respect for human rights. Poverty, he argued, was as formidable an enemy to human rights as the most repressive dictatorship. "These rights are hollow for any individual who starves to death," he explained. The United States must reject the short-term safety of the status quo for the longer-term promise of constructive change. "We need not fear change if we build into it more equity and more participation. Indeed, fear of social change is the thing we need to fear the most. If we are afraid of it and try to preserve that which is eroding beneath our feet, we will fail." The world was in the process of a great transformation. "I

believe that we are at end of the period of cold wars, in the middle of the
era of detente, and just beginning to find ways to build the structures of
cooperation." The new era would take some getting used to. "Cooperation
will demand a different substance and a different style than confrontation." But the challenge included an opportunity. "I believe that cooperation for the common good of humankind can be as powerful an incentive to
our imagination as fear for our survival." Americans had little alternative.
"Cooperation for the common good, for the protection and promotion of
human rights, is the way to survival."[34]

Young sometimes sparked as much consternation within the Carter
camp as among Carter's conservative critics. He eventually was fired after
talking to representatives of the Palestine Liberation Organization. But
others in the administration held many of the same views. Cyrus Vance
asserted in a keynote address on the administration's Africa policy that the
United States must avoid treating Africa "as a testing ground of East-West
competition." Vance continued, "The most effective policies toward Africa are affirmative policies. They should not be reactive to what other
powers do, nor to crises as they arise. Daily headlines should not set our
agenda for progress. A negative, reactive American policy that seeks only
to oppose Soviet or Cuban involvement in Africa would be dangerous and
futile. Our best course is to help resolve the problems which create opportunities for external intervention."[35]

Beyond stressing human rights and North-South issues, Carter spent a
great deal of time on another matter—energy—that had figured in the
Cold War, but had lately acquired an importance having little to do with
the rivalry between the United States and the Soviet Union. The Arab oil
embargo of 1973 and the firming up of the Organization of Petroleum
Exporting Countries during the mid-1970s placed the American economy,
and the economies of other oil-importing states, in potentially grave jeopardy. Already the huge price rises OPEC directed had drained dollars out
of America, Europe, and Japan, and were threatening the stability of the
world monetary framework. What combination of inflation and recession
in the industrial importing countries the transfer of wealth to the exporters
would produce remained to be seen, but the prognosis wasn't promising.

Consequently, Carter placed the energy crisis near the top of his list of
priorities. "With the exception of preventing war," he told a national
television audience during his third month in office, "this is the greatest
challenge that our country will face during our lifetime." The energy
squeeze would test the character of the American people, and the ability of
Congress and the president to govern. It would demand of the United
States an unprecedented effort. "This difficult effort will be the 'moral
equivalent of war,' except that we will be uniting our efforts to build and
not to destroy." Unless the country acted quickly to reduce dependence on
imported oil, the economic, political, and social costs would be exorbitant.
"Now we have a choice. But if we wait we will constantly live in fear of

embargoes. We could endanger our freedom as a sovereign nation to act in foreign affairs. Within ten years we would not be able to import enough oil from any country, at any acceptable price. . . . Inflation will soar; production will go down; people will lose their jobs. Intense competition for oil will build up among nations and also among the different regions within our own country. This has already started."

Carter outlined a multi-step program for diminishing America's dependence on foreign supplies of energy. He described measures to create a strategic petroleum reserve, to increase domestic oil production, to develop alternative sources of energy, and to encourage conservation. Most significantly, he called on the American people to reduce their expectations of the future, and to prepare to live with a certain level of discomfort. "I'm sure that each of you will find something you don't like about the specifics of our proposal. It will demand that we make sacrifices and changes in every life. To some degree, the sacrifices will be painful." But meaningful sacrifices were always painful, and this one was both meaningful and necessary.[36]

CHAPTER SIX

Old Verities Die Hardest
1977–1984

Paul-Revering the nuclear age

Asking for sacrifice in the name of petroleum self-sufficiency was Carter's first mistake. The American political system is reasonably good at eliciting public sacrifice when the sacrifice seems directly related to national security and honor, and when it promises not to last too long. In both world wars, Americans complained hardly at all about throwing their lives and fortunes into the breach. They complained more about the Korean War, which seemed considerably less linked to essential American interests, but that war ended before the complaints achieved critical mass. Vietnam suffered from the double disability of being both marginal to American security and overly long, and eventually Americans demonstrated that they wouldn't put up with it any more. Carter could talk all he wanted about the energy shortage being the moral equivalent of war, but talk didn't make it so. American presidents have rarely succeeded as scolds. Given a choice, American voters have generally preferred being told what they *can* have, to being told what they *can't*.

Once having declared moral war on the energy problem, Carter needed to show progress if he expected popular support. But the news from the front only got worse. The Iranian revolution of 1979 re-created the service-station lines of 1973 and ratcheted prices up further. At a loss as to what else to do, Carter fired several of his generals—cabinet members, that is—and went on television once more to rally the rank and file. Again he called for sacrifice, and again he asserted that the struggle would be protracted and trying. America, he said, faced a "crisis of the spirit," which could be cured only by a new dedication and a new sense of purpose. Victory would come, but slowly. "Little by little, we can and we must rebuild our confidence."[1]

In a tactical political sense, Carter committed a fatal error in conflating objective problems, in this case high levels of oil imports, with subjective questions such as national self-confidence. It was natural for him to try to shift the ground of public debate into the realm of the intangible, since the tangible world was proving so intractable. But in doing so, he opened himself to the attacks of opponents who promised to restore self-confidence without demanding the sacrifices Carter called for.

Of opponents, Carter had plenty. Some he developed on his own. Others he inherited. Most formidable among the latter were a group of Republicans and conservative Democrats who had never reconciled themselves to detente and the de-ideologizing of the Cold War. Senator Henry Jackson raised perhaps the greatest ruckus. The Washington Democrat had for years supported the dry-powder school of strategic thinking. In 1951, hoping to graduate to the Senate from the House of Representatives, which he found confining after five terms, Jackson had advocated a huge buildup of American military strength, well beyond the levels already in the works under the double impetus of NSC 68 and the Korean War. Following a Soviet atomic test that showed the Russians diminishing America's weapons edge, he declared that the United States must "go all out" to preserve its technological lead. He recommended increasing the portion of the federal budget devoted to atomic energy, including weapons research, by as much as ten times.

In advocating such policies, Jackson made plain his visceral opposition to communism and all that the communists stood for. But he also demonstrated an awareness that bad relations between the United States and the Soviet Union meant good business for the defense contractors of his home Puget Sound region. During his campaign for the Senate, he asserted that a tough policy against the communists would result in the area becoming "one of the principal arsenals of democracy for the United States, just as it was in the last war." He added, "Many plants will be converting from peace time to all-out defense production."[2]

Jackson's campaign strategy worked, landing him a Senate seat. During the next twenty years, his actions in the upper house showed this same combination of personal ambition, ideological conviction, and political calculation. He ran for re-election in 1958 on a platform of strengthening both American air power and the economy of Washington state. "The Air Force needs a fleet of 1800 Boeing Airplane Company jet bombers and tankers," he declared, brand-specifically. During the 1960s, while still solicitous of the welfare of the largest private employer in his state, he shifted his emphasis to nuclear-missile submarines. He became an ardent supporter of the Trident submarine, helping guide the program through the Senate, and succeeding in basing the big U-boats in Bangor, Washington—to the tune of half a billion dollars in construction costs, and some one hundred million per year in federal payroll expenditures.[3]

In the early 1970s, Jackson neatly tied his Cold War fervor to another passion: Israel and the condition of world Jewry, especially that portion living in the Soviet Union. He introduced an amendment to the 1970 defense procurement bill, authorizing the president to provide Israel with whatever weapons Israel's security required. Jackson justified the measure by warning of "Soviet ambition in the Middle East," and declaring that "the front line of Western defense in the Middle East is manned by the men and women of Israel." Despite the opposition of the Senate's senior Democrat on foreign affairs, William Fulbright, the amendment passed overwhelmingly. Shortly thereafter, Jackson advocated linking liberalization of U.S.-Soviet trade relations—specifically, the Nixon-Kissinger goal of most-favored-nation status for the Soviets—to the Kremlin's loosening of strictures on Jewish emigration. He wrote the linkage into an amendment to a bill regulating U.S.-Soviet trade. This amendment proved politically irresistible, and nearly three-quarters of the Senate joined him as cosponsors. The House demonstrated equal enthusiasm for a similar measure introduced by Charles Vanik. Though the Nixon administration deplored the idea of muddying the waters of detente with the extraneous issue of the Kremlin's emigration policy, there was little the administration could do in the face of such broad-gauged backing. The trade bill to which the Jackson-Vanik amendment was attached involved enough divergent interests to require many months to gain congressional approval. By the time it did, at the end of 1974, Nixon had been succeeded by Ford, who was even less able to resist the pro-amendment pressure. Ford signed the bill, with amendment, into law in January 1975.[4]

For Jackson, the connecting of U.S.-Soviet trade to the fate of Soviet Jews was a winner either way Moscow responded. If the Kremlin opened the doors to emigration, the Jews would get out. This would be good for them personally, and good for Jackson politically. Even before the Jackson-Vanik amendment became law, Rabbi Mitchell Wohlberg of Beth Sholom synagogue in Washington, D.C., spoke for many American Jews when he asserted, "We have a hero and his name is Henry Jackson. More than any other leader on the American political scene, Senator Henry Jackson has been there in our time of troubles and pain." Should the troubles and pain lessen as a result of Jackson's legislative efforts, he would certainly be remembered. On the other hand, if the Kremlin rejected the American condition as meddling in internal Soviet affairs, Moscow would forfeit most-favored-nation status. This would be bad for detente, and again good for Jackson. Having staked his claim to the Cold War, it was in Jackson's interest to do what he could to prevent whatever mitigated tension between the superpowers.

In the event, the Soviets rejected the Jackson-Vanick condition. Detente lost ground, and Jackson and other Cold Warriors gained it.

Jackson may or may not have recognized in advance the full potential of combining opposition to detente with support for Israel and other issues of concern to American Jews, but the combination was brilliantly successful for several years. To a certain degree, American enthusiasm for Israel— not simply among American Jews—was inversely related to Americans' confidence in their own country. After the Vietnam venture turned bad, at a time when many Americans feared that the United States lacked the necessary combination of capacity and will to successfully defend American interests in an unfriendly world, beset but intrepid Israel served as a surrogate winner. Americans like underdogs, especially those who beat the bullies and come out on top. Israel fit the description perfectly. The 1973 Yom Kippur War, in which Israel absorbed a surprise attack, then proceeded to trounce its attackers, elicited strong American support for Israel, despite the fact that the war produced the closest military brush between the United States and the Soviet Union since the Cuban missile crisis, and despite the fact that American support for Israel triggered the Arab oil embargo that led to the gas lines and all the economic and psychological distress the petroleum shortfall produced. The 1976 Israeli raid on Entebbe epitomized in the minds of many Americans what the United States ought to be able to do, and perhaps once *had* been able to do, but could do no longer.

Moreover, Israel's predicament appealed to the same desire for clarity in international affairs that had characterized American thinking about the Cold War. For the Israelis, surrounded by sworn enemies who refused to grant Israel even the minimal respect shown by an acknowledgment of Israel's right to exist, the essential issues were undeniable and unconfused. Strength was the key to survival. Irresolution would bring destruction. Cold Warriors in the United States could only wish that the issues confronting America were so clear.

For such reasons, and doubtless for others that varied from individual to individual, the attack on detente emerged strongly from quarters devoted to the well-being of Israel. Spearheading the attack was *Commentary* magazine, the organ of the American Jewish Committee. Aiming the spearhead was Norman Podhoretz, *Commentary*'s incendiary editor. Podhoretz, like other neoconservatives, arrived at his position on the right wing of American politics following a flight from the left. During the 1930s, many American intellectuals had viewed socialism as a promising alternative to America's apparently moribund capitalism. Americans Jews, at that time as at others represented among intellectuals in larger numbers than their proportion of the American population would have indicated, had an additional reason for looking hopefully on socialism. Stalin seemed more interested than other world leaders, including Franklin Roosevelt, in standing up to Hitler, whose depredations against Jews increased by the month. The disillusionment of the left began with the Soviet-German non-aggression pact of 1939, which canceled most of the credit the Soviet Union had accumulated on the anti-fascist score. The disillusionment continued through World War II, whose enormous destructiveness, together with the previously inconceivable horror of the Nazi holocaust, shattered nearly everyone's belief in the essential goodness of human nature, on which faith in socialism ultimately rested. By the early 1950s, a sizable group of former radicals had joined the ranks of what Podhoretz called "hard anti-Communism." Podhoretz and like-thinkers such as Irving Kristol and Nathan Glazer provided intellectual legitimacy to the Mc-Carthyist movement, defending the Wisconsin senator's motives and aims, if not all his methods.[5]

The 1960s unsettled the newly conservative intellectuals as much as it did the rest of American society, and cast them in a dozen directions. But by the early 1970s, they were regrouping around the twin issues of support for Israel and opposition to detente. *Commentary*, under Podhoretz's guidance, provided the most visible forum for their opinions. Theodore Draper, a consistent critic—from the left—of Lyndon Johnson during the 1960s, published a series of articles in *Commentary* denouncing detente as "an unmitigated snare and delusion" and describing the Nixon-Kissinger policy as the new appeasement. Detente, Draper said, consisted chiefly of American give and Soviet take. "Appeasement was built into detente whenever we adapted ourselves to them but they did not adapt themselves to us." Unless Americans and other complacent Westerners woke up to what the Soviets were doing in detente's name, the consequences would be "cumulatively disastrous."[6]

Podhoretz opened the pages of *Commentary* to others writing in a similar vein, but the sharpest cuts he delivered himself. In a 1976 article entitled "Making the World Safe for Communism," Podhoretz castigated Kissinger (Nixon now being under post-presidential house arrest in San Clemente) for the worst form of appeasing hypocrisy. The secretary of state, Podhoretz wrote, "often sounds like Churchill and just as often acts like Chamberlain." Podhoretz accused Kissinger of promoting a new and insidious form of isolationism: what Podhoretz labeled "Finlandization from within." The future looked bleak. "If it should turn out that the new isolationism has indeed triumphed among the people as completely as it has among the elites, then the United States will celebrate its two-hundredth birthday by betraying the heritage of liberty which has earned it the wonder and envy of the world from the moment of its founding to this, and by helping to make that world safe for the most determined and ferocious and barbarous enemies of liberty ever to have appeared on the earth."[7]

For all their vitriolic talents, Podhoretz and the other neoconservatives suffered from the handicap of being intellectuals in a society that distrusts intellectuals. To add weight to their arguments, they required someone from beyond the intellectual ranks. Henry Jackson might have served, but Jackson was a career politician, a member of a class rated by most of his compatriots as being even lower than the intellectuals. What the detente-busters really needed was someone who could speak about the issues of superpower relations with the authority of experience, and who could attack detente from inside the system.

They got just what they were looking for in Paul Nitze. Nitze, of NSC

68 and the Gaither report, had subsequently served as navy secretary and then deputy secretary of defense during the Democratic administrations of the 1960s. Nixon and Kissinger appreciated the trouble Nitze could cause if left outside the policy process, so they attempted to co-opt him by bringing him aboard. At first, Nixon named Nitze to be the American ambassador to West Germany, a post sufficiently prestigious to stroke Nitze's ego and sufficiently distant to keep him out of the administration's hair. But the Bonn appointment stalled in the Senate, where Fulbright thought the nominee too conservative and Goldwater thought him too liberal. The administration next considered sending Nitze to the embassy in Tokyo, which was even farther away than Bonn and just as prestigious. But this idea failed for the same reasons as the first. Nixon finally succeeded on the third try: appointment of Nitze as representative of the secretary of defense at the SALT negotiations soon to begin at Helsinki.

Nitze proved poorly suited to his new job. He had never lost the basic convictions that informed NSC 68 and the Gaither report. He still held that the Soviet Union was a dangerous and implacable foe, and that the appropriate way for America to deal with this foe was from a position of military superiority. In 1963, Nitze had told the armed services committee of the Senate that he was "a strong believer in the importance of maintaining superiority over the Communist bloc in every element of our military power." During the SALT negotiations, he became aware that the Nixon administration had given up on superiority in favor of parity, and he found the concession bitterly distasteful. He might have stayed with the administration, the better to diminish the damage, had Kissinger not insisted on bypassing the official SALT team in favor of back-channel talks direct to the Kremlin. In the final stages of the negotiations leading to the May 1972 summit, Nitze and the others on the SALT team were left almost completely in the dark. At the summit, Nitze suffered the personal indignity of being stranded at the Moscow airport, assigned a Russian driver who spoke no English, and detained by security guards at the American embassy, requiring rescue by a Soviet diplomat who recognized him when no American did. Nitze hung on gamely as the Watergate scandal unfolded, but in the spring of 1974 he quit in a public huff. Via press release, he declared that after nearly three decades of working on national defense issues, and after nearly five years of working on SALT, he now saw "little prospect" of progress toward measures that would enhance American security. He told a congressional committee that the Soviet Union was on its way to gaining strategic superiority over the United States, and that the Russians would accomplish this task unless Americans pulled their heads out of the sand. When Nitze's statement provoked Kissinger to reply, "What in the name of God *is* strategic superiority? What is the significance of it, politically, militarily, operationally, at these levels of numbers? What do you do with it?," Nitze called Kissinger "a traitor to his country."[8]

No longer yoked to the Republican administration, Nitze felt free to foil the detentist designs of the traitorous Kissinger. He fashioned his arguments against SALT into an article that appeared in the January 1976 issue of *Foreign Affairs*. The timing of the article—the beginning of a presidential election year—and the authorship—by a professedly eye-opened former SALT negotiator—were as significant as its content. But the content was substantial. Nitze declared that the SALT process, as currently practiced by Kissinger and the Ford administration, would lead to disaster. "There is every prospect that under the terms of the SALT agreements the Soviet Union will continue to pursue a nuclear superiority that is not merely quantitative but designed to produce a theoretical war-winning capability. Further, there is a major risk that, if such a condition were achieved, the Soviet Union would adjust its policies and actions in ways that would undermine the present detente situation, with results that could only resurrect the danger of nuclear confrontation or, alternatively, increase the prospect of Soviet expansion through other means of pressure." In these circumstances, a buildup of American forces, rather than their limitation, was necessary. "If and only if the United States now takes action to redress the impending strategic imbalance can the Soviet Union be persuaded to abandon its quest for superiority and to resume the path of meaningful limitations and reductions through negotiation."

Not satisfied with bashing arms control, Nitze went after detente hammer and tongs. In less inflammatory language than Draper, Podhoretz, and the rest of the *Commentary* crowd were using, Nitze drove home essentially the same point: that detente was a one-sided fraud in which the United States made all the concessions and Moscow all the gains. Recent events in Southeast Asia, the Middle East, and elsewhere offered not the slightest evidence of Soviet mellowing. The rhetoric may have changed, but the actions remained as before. "In sum total, there are strong grounds for concluding that in Soviet eyes 'detente' is not that different from what we used to call the 'Cold War.' "

Nitze had contributed, in the Gaither report, to the popular belief in a "missile gap," and he remembered the effectiveness of that notion in breaking down Eisenhower's resistance to increased defense spending. Now he deployed a slightly more esoteric, but in the end equally effective device: a "throw-weight gap." Unlike the United States, which had spread its nuclear weapons roughly evenly among the three legs—land, sea, air—of its "triad" (reflecting both the rivalry among the military services and the fact that the United States had easy access to the sea, as well as allies willing to accept forward-based American aircraft), the Soviet Union leaned heavily on land-based missiles. Partly to compensate for the lesser accuracy of these missiles, compared with American missiles, and partly to take advantage of its engineers' expertise in building big rockets, the Kremlin loaded its land missiles with larger warheads than America's generally carried. This gave Moscow an advantage in "throw-weight"— an advantage that Nitze and the anti-detentists played to *their* advantage.

In his *Foreign Affairs* article, Nitze produced a pair of graphs demonstrating (after the fashion of the Gaither report) what he deemed to be ominous trends in throw-weight. At the time of the Cuban crisis, the graphs showed, the United States had enjoyed a sizable edge in throw-weight over the Russians. The Kremlin had caught up during the late 1960s. Even now, the Soviet building program continued apace. The balance was tilting more steeply against America with each passing year. More alarming still was that, due to the configuration and differential protection of the nuclear forces of each side, the Soviets' advantage in throw-weight would increase in the event of what seemed the most likely form of limited nuclear exchange: one directed at the two countries' forces rather than their populations.

In other words, Nitze explained, Moscow could expect to emerge from a first round of nuclear war relatively stronger than it entered. This danger to America would worsen in the future also, notwithstanding America's new China policy and the support of the NATO allies. "By 1977, after a Soviet-initiated counterforce strike against the United States to which the United States responded with a counterforce strike, the Soviet Union would have remaining forces sufficient to destroy Chinese and European NATO nuclear capability, attack U.S. population and conventional military targets, and still have a remaining force throw-weight in excess of that of the United States. And after 1977 the Soviet advantage after the assumed attack mounts rapidly."[9]

What Nitze described would gain wider currency as the "window of vulnerability," and would spur the weapons-building boom of the Reagan era. For the present, though, Reagan and Nitze were still on the outside. The Ford administration would have been happy to keep them there. But Reagan was running for president, and the noise Nitze was making lent plausibility to Reagan's complaints against detente. In an effort to silence Nitze, and therefore sidetrack Reagan, the administration chose to bring Nitze back inside. During the summer of 1976, George Bush, Ford's CIA director, created a shadow intelligence board known as Team B. Nitze joined the group, as did several other detente-distrusters who later took positions in the Reagan administration. The exercise achieved its purpose of stymieing Reagan, although, by further discrediting detente, it did little for Ford's chances in the November contest against Carter.

The details of a report Team B produced in December 1976 were closely held, yet the gist of the report soon spilled. A professional intelligence officer who saw the report remarked, "It was more than somber—it was very grim. It flatly states the judgment that the Soviet Union is seeking superiority over the United States forces." One B-teamer, General George Keegan, put the matter in stronger terms. "I am unaware," Keegan said, "of a single important category in which the Soviets have not established a significant lead over the United States." William Van Cleave, another member, declared, "I think it's getting to the point that if we can make a trade with the Soviet Union of defense establishments, I'd be heartily in favor of it."[10]

Needless to say, such remarks were intended to render it difficult for the incoming Carter administration to continue down the path of detente. Working toward the same obstructive goal was the Committee on the Present Danger. The CPD, which took the name of a hard-line anti-Soviet organization of the early 1950s, included several B-teamers, Nitze among them, as well as Podhoretz and a broad array of conservatives, neo-conservatives, Cold War liberals, former government officials and military officers, defense-industry executives, financiers, labor-union representatives, university professors, and the odd novelist and poet. It dedicated itself to waking America to the danger its members believed America faced from Soviet ambitions—in other words, to reviving the Cold War. In the present-dangerists' view, the Cold War had never ended, only the American side of it. Brezhnev might not be Stalin—although the Czechs could be excused, since the 1968 Soviet invasion, for missing the difference—but the personnel in the Kremlin had always mattered less than the ideology of the Soviet system. That ideology was intrinsically expansionist, and would brook no peace with other ideologies and systems until it had eliminated or absorbed them. American leaders had deluded themselves, and had tried to delude the American people, into thinking ideology didn't matter. They couldn't be more wrong. Unless change came soon, American security and world peace would suffer gravely for their mistake.

Detente undone

The drumbeat of criticism mounted as Carter entered office, though for a time the Democratic president managed to ignore most of it. His rejection of an "inordinate fear of communism" convicted him, in present-dangerist minds, of stunning naïveté, and his preachings about human rights ignored the greatest danger to human rights of all: the prospect of a world communist state. During 1977 and 1978, Carter blunted the criticism with diplomatic progress in other areas, notably the Middle East. Carter's unrelenting and ultimately successful efforts to mediate a peace accord between Israel and Egypt appeared to prove, at least to those willing to be convinced, that there did indeed exist important problems that could be resolved if American leaders allowed themselves to look beyond the conceptual confines of the struggle against communism.

But in 1979 Carter's position disintegrated. The year began with the flight of Shah Pahlavi from Iran and the return of Ayatollah Khomeini from exile. By itself, the Iranian revolution should have had little effect on the debate in the United States over the nature of the communist threat, since Khomeini and the mullahs who now ruled Iran evinced no more affinity for the atheists of the East than for the Great Satan of the West. Yet the loss of a long-time ally in a crucial region of the world, together with Carter's vacillation in determining how to deal with the shah's collapse, reflected poorly on the administration's judgment. Americans waiting in gas lines to pay higher-than-ever prices had plenty of time to reflect on precisely this matter.

The summer of 1979 brought the overthrow of the Somoza regime in Nicaragua. The Nicaraguan revolution seemed more closely related to the ideological contest between the United States and the Soviet Union, since the Sandinistas who controlled the successor government unabashedly took their inspiration from Castro's Cuba. Carter tried to avoid the mistakes he thought the Eisenhower administration had made with Castro, and he kept sending American aid. But the Sandinistas hadn't mounted their revolution to become bourgeois reformers. They took the Americans' food and money, and went about the business of remaking Nicaraguan politics and society along radical lines.

The proponents of a revived Cold War capitalized on Carter's discomfiture over Iran and Nicaragua. In November 1979, *Commentary* carried a withering assault on the administration's policies by CPD member and political scientist Jeane Kirkpatrick. Under the title "Dictators and Double Standards," Kirkpatrick damned Carter for losing Iran and Nicaragua, much as conservatives thirty years earlier had damned Truman for losing China. Kirkpatrick asserted that Carter's policies toward Pahlavi and Somoza, two rulers who, though less than democratic, were better than what was certain to follow, could hardly have been more perverse. "In each country," Kirkpatrick asserted, "the Carter administration not only failed to prevent the undesired outcome, it actively collaborated in the replacement of moderate autocrats friendly to American interests with less friendly autocrats of extremist persuasion."

Kirkpatrick drew what she deemed a crucial distinction between authoritarian autocracies of the right, like those of Somoza and Pahlavi, and totalitarian autocracies of the left, like those found in communist countries. The former allowed the continued existence of traditional social institutions, which served to shelter individuals from the oppressive power of the state. The latter tolerated no interference with state power. On account of this difference, the former offered greater hope for liberalization—which was to say, some hope as opposed to none. "Although there is no instance of a revolutionary 'socialist' or Communist society being democratized," Kirkpatrick wrote, "right-wing autocracies do sometimes evolve into democracies—given time, propitious economic, social and political circumstances, talented leaders, and a strong indigenous demand for representative government."

Kirkpatrick contended that the Carter administration employed a double standard in dealing with dictators. In stressing the need to move beyond the Cold War, the administration displayed an unsettling tendency to destabilize American allies, while leaving communist dictatorships firmly in place. "The American commitment to 'change' in the abstract ends up by aligning us tacitly with Soviet clients and irresponsible extremists like the Ayatollah Khomeini or, in the end, Yasir Arafat." (With other neoconservatives, Kirkpatrick considered Israel a strong ally against radicalism.) She continued, "So far, assisting 'change' has not led the Carter administration to undertake the destabilization of a *Communist* country. The principles of self-determination and nonintervention are thus both selectively applied. We seem to accept the status quo in Communist nations (in the name of 'diversity' and national autonomy), but not in nations ruled by 'right-wing' dictators or white oligarchies."

The Democrats practiced not mere inconsistencies, Kirkpatrick said, but dangerously foolish inconsistencies that undercut America's strategic interests. "What makes the inconsistencies of the Carter administration so noteworthy are, first, the administration's moralism—which renders it especially vulnerable to charges of hypocrisy; and, second, the administration's predilection for policies that violate the strategic and economic interests of the United States. The administration's conception of national interest borders on doublethink: it finds friendly powers to be guilty representatives of the status quo and views the triumph of unfriendly groups as beneficial to America's 'true interests.' "

Kirkpatrick decried the Democrats' tendency to blame America for the troubles of the world. If America suffered a crisis of the spirit, she said, it was a crisis inflicted by liberals. Americans needed to remember what was right with the United States. As they did so, they would learn to tell foes from friends. If foreigners described America as colonialist, genocidal, expansionist, and racist, they weren't friends. The United States wasn't a colonialist power. It didn't engage in genocide. It didn't force itself on other people. Though its record on race wasn't perfect, it had "moved further, faster, in eliminating domestic racism than any multiracial society in the world or in history." In dealing with developing nations, the United States must reject the posture of "continuous self-abasement and apology" that the Carter administration had foisted on the American people. The liberals didn't have a monopoly on idealism, or on genuine liberalism, for that matter. Indeed, they egregiously misunderstood the meaning of the terms. "Liberal idealism need not be identical with masochism, and need not be incompatible with the defense of freedom and the national interest."[11]

Kirkpatrick's article caught the attention of at least one person whose opinion mattered—or would matter soon. Ronald Reagan read the piece, and decided that its author would fit well into a Reagan administration. As things turned out, Kirkpatrick got a job as Reagan's representative to the United Nations, and the distinction between redeemable authoritarians and irredeemable totalitarians provided the basis for much American foreign policy during the 1980s.

While Reagan was trying to figure out what job to give Kirkpatrick, another pair of events significantly enhanced his prospects of becoming the nation's top job-giver. In the process, they contributed to the resurrection of the Cold War. The seizure of several dozen hostages from the American embassy in Tehran, though as unrelated to the superpower rivalry as the rest of the Iranian revolution, seemed further evidence of American impotence in the face of global challenge. Carter handled the matter reasonably well from a humanitarian standpoint, and the fact that all the hostages eventually returned unharmed resulted largely from his refusal—with the glaring exception of a botched rescue mission—to do anything rash. From a political standpoint, though, he could hardly have handled the matter worse. By making the hostages the center of his attention, he paralyzed his administration. He gave the hostage-holders less incentive than they might have had to release them, and afforded his domestic opponents an easy target. With reason, many Americans considered intolerable a situation in which the United States found itself helpless in the face of extreme provocation by the medieval-minded government of a third-rate country. Increasingly, they heeded those who promised to restore American dignity and power.

Detente's coup de grace came in December 1979, when the Soviet Union invaded Afghanistan. Moscow's move probably followed less from a grand design upon the Persian Gulf or the Indian Ocean than from a desire to keep the Islamic fundamentalism that had taken control of Iran from doing the same to Afghanistan and sweeping into the Muslim republics of the Soviet Union. All the same, Carter chose to interpret the move as a direct challenge. Whatever reality underlay the tumultuous situation in Middle East, political circumstances in the United States left him almost no choice. Republicans continued to rebuke him for ineptitude and flaccidity in dealing with Iran and Nicaragua. And increasing numbers of Democrats were following suit. The SALT II treaty, signed by Carter and Brezhnev in Vienna in June 1979, had stalled in the Senate. Henry Jackson, showing no more respect for detente as pursued by a president of his own party than he had for the Republican version, described Carter's policy as "appeasement in its purest form." In August, another Democrat, Senator Church, had announced the discovery of a Soviet "combat brigade" in Cuba. The unit in question had in fact been in Cuba for several years at least, but Church hadn't been facing a tough re-election campaign in conservative Idaho then, and no one had paid attention. Now everyone paid attention. In responding to Church's announcement, the Carter administration attempted the delicate task of demonstrating its determination not to allow a threat to develop in Cuba, and at the same time reassuring Americans that the matter was nothing to worry about. It failed on both counts. After declaring the status quo unacceptable, it ended up leaving matters essentially as they stood.[12]

Within the administration itself, hardliners were gaining the ascendancy. For nearly three years, the moderate Vance had tried to offset the reactive anti-Sovietism of Carter's national security adviser, Zbigniew

Brzezinski. For most of that period, Carter had listened to both men. But the events of 1979—as they occurred overseas, and as they were interpreted at home—pushed the president irrevocably toward Brzezinski. With an impolitic ingenuousness that only made his problems worse, Carter declared, on the final day of 1979, "My opinion of the Russians has changed more drastically in the last week than in the two and one-half years before that." Not to be mistaken, Carter repeated himself: "The action of the Soviets has made a more dramatic change in my opinion of what the Soviets' ultimate goals are than anything they've done in the previous time that I've been in office."[13]

In two sentences, Carter conceded victory to the anti-detentists. He confessed to the naïveté they had been charging him with since 1976, and he implicitly endorsed their call for a return to confrontation. He made the endorsement explicit during the next month. He withdrew the SALT II treaty from Senate consideration. He embargoed grain exports to the Soviet Union, beyond those already committed. He barred the sale of high-technology products to the Soviets. He suspended Russian fishing privileges in American waters. He canceled a variety of academic and cultural exchange programs. He ordered the withdrawal of the American team from the 1980 summer Olympic games in Moscow.

And in his State of the Union speech in late January 1980, he set forth what came to be called the Carter Doctrine. After recapitulating the critical moments of the Cold War, with emphasis on the measures his predecessors had taken to block Soviet expansionism, Carter placed himself squarely in their footsteps. The Soviet invasion of Afghanistan, he explained, had brought Moscow's forces "to within 300 miles of the Indian Ocean and close to the Straits of Hormuz—a waterway through which most of the world's oil must flow." Russian entrenchment in Afghanistan would pose "a grave threat" to the flow of Middle Eastern oil. In the most direct language possible, Carter warned the Kremlin against further advances. "Let our position be absolutely clear. An attempt by any outside force to gain control of the Persian Gulf region will be regarded as an assault on the vital interests of the United States of America, and such an assault will be repelled by any means necessary, including military force." To add weight to this last statement, Carter conspicuously announced an increase for defense in the federal budget he was proposing for fiscal 1981.[14]

The 5 percent real increase Carter requested for 1981, together with equal increases he projected for each of the four succeeding years, didn't match, in relative terms, the nearly 200 percent increase NSC 68 had called for three decades before, but in dollar terms it did. And together with Carter's adoption of a Trumanesque posture in the Persian Gulf, it brought the curtain clattering down on the era of detente. At this point, the only question that remained was how cold the new Cold War would get.

Actually, there *was* another question, but one that would be of greater interest to historians and other after-the-facters than it was to Carter, the present-dangerists, and those struggling to make policy at the time. This question involved causation, to wit: Who killed detente? For Carter and the rest in 1980, it was sufficient that detente was dead. They could leave the autopsy to those without care for hostages, oil supplies, and elections.

The obvious answer was that the Soviets killed detente, chiefly by their invasion of Afghanistan. By blatantly violating the principles of restraint that were supposed to have governed superpower relations since the Moscow summit of 1972, the Kremlin left the United States no choice but to retaliate. Carter exaggerated when he called the Soviet invasion of Afghanistan "the greatest threat to peace since the Second World War." He must have been forgetting Korea and Vietnam, which not only threatened but shattered peace, and the Cuban missile crisis. But the president's overall response wasn't out of line with the provocation.

Obvious answers usually contain some truth, and this one about the Soviet role in detente's demise is no exception. Yet if the Soviets hadn't dispatched detente in December 1979, it probably wouldn't have survived the American election campaign of 1980. No viable Republican presidential candidate was about to go anywhere near detente, and, considering the condition of Carter's political fortunes, the president almost certainly would have been forced to back away from what remained of the policy. The SALT II treaty had little chance of gaining Senate approval even before the Afghanistan affair—which, in fact, gave the Carter administration an excuse to pull the treaty and blame the Russians. As Brzezinski commented in urging such a strategy on the president, "If we blame the Soviet invasion of Afghanistan for the delay of SALT, it will be less a political setback for us."[15]

Although it would be a mistake (albeit a common one, for Americans) to attribute too much influence to American actions in determining what the Soviet Union did, the Soviets understood American politics well enough to realize that detente probably had only a few months left to live. They certainly didn't go into Afghanistan *because of* the beating detente was taking in the United States, but, to the extent they considered the likely American reaction to an invasion, they must have calculated that they had little to lose on detente's score. Afghanistan or not, the Kremlin faced a rough time in America in the upcoming election campaign. A miracle might save Carter, but communists aren't supposed to believe in miracles. In any case, his salvation might well have been at Moscow's expense. A Republican victory promised even less.

Fully as much as the Soviets, the present-dangerists were responsible for the death of detente. They didn't deny the charge. On the contrary, they embraced it. Since the early 1970s, the likes of Jackson, Podhoretz, Nitze, and Kirkpatrick had opposed any relaxation between the superpowers, and their opposition had only increased as the decade aged. Taking their cue from Winston Churchill's attacks on appeasement during the

1930s, they strove mightily to destroy detente forty years later, and they considered themselves patriots for their efforts.

Why did they succeed? For a number of reasons. In the first place, the Soviets cooperated in acting the villain. The invasion of Afghanistan was the clearest case of villainy, but from the beginning of detente, the Kremlin had persisted in activities that sorely tested the 1972 understanding that the superpowers should avoid actions that disrupted international affairs. In Angola and in the Horn of Africa (where the Russians backed leftist coup-makers in Ethiopia and Somalia), directly and through Cuban proxies, the Soviet Union continued to play the Cold War game of beggar-thy-superpower-neighbor. During the 1973 Middle East war, Moscow had been so far from cooperating in calming matters that the Nixon White House—that den of detentists—had felt required to place American military forces on alert to warn the Kremlin away from intervening on Egypt's side. Needless to say, the Soviets believed they had justification for acting as they did, but even detente's supporters in America often found the Kremlin's arguments flimsy.

In the second place, the attack on detente succeeded because detente's proponents oversold their product. Most Americans had never gotten used to the fact that friction exists between great powers even in the best of times. Recalling, historically if not personally, the days of America's relative isolation in the nineteenth century, when the United States had had much of a hemisphere more or less to itself, Americans found it difficult to adjust to the shrunken-globe world politics of the twentieth century. (This was one reason why the Cold War had come as a surprise to many Americans in the late 1940s.) When Nixon went to China and the Soviet Union in 1972, laughing with and toasting Mao and Brezhnev, and when the president spoke of a new era in international relations, it was easy for Americans to gain the impression that affairs among the great powers would be as friendly as the meetings of their leaders seemed to be. In Beijing in 1972 Nixon asserted, "This was the week that changed the world." He spoke too optimistically, and his excess optimism came back to cost detente.[16]

In the third place, detente died because the Carter administration did a dismal job directing it. Through his first three years in office, Carter failed to resolve the difference of opinion that separated the Brzezinski camp in his administration from the Vance group. The result of this failure was a policy that sometimes exhibited a split personality. One notable address by Carter at the United States Naval Academy in June 1978 left everyone puzzling over what the administration was up to. "We must avoid excessive swings in the public mood in our country, from euphoria when things are going well to despair when they are not, from an exaggerated sense of compatibility with the Soviet Union to open expressions of hostility," the president declared. Then he proceeded to demonstrate just such a swing. "Detente between our two countries is essential to world peace," he said, before adding, "To the Soviet Union, detente seems to mean a continuing aggressive struggle for political advantage and increased influence." The latter statement wasn't much of a recommendation for something defined as essential to world peace. Carter didn't help detente's case when he presented it to Moscow as a take-it-or-leave-it proposition. "The Soviet Union can choose either confrontation or cooperation. The United States is adequately prepared to meet either choice."[17]

The Soviets responded by declaring Carter's speech "strange." Washington reporters, despite readier access to administration officials, had no better luck figuring out what was going on. The *Washington Post* subtitled its account of the address "Two Different Speeches." The *New York Times* said the speech seemed "contradictory." The *Los Angeles Times*'s diplomatic correspondent called the message "as ambiguous as the conflicting policies and complex circumstances from which it sprang." Syndicated columnist Mary McGrory asked, "What was Carter saying?" and concluded, "All he said was that he doesn't even know what he thinks."[18]

Cold War II

But perhaps the most important reason for detente's demise was that Americans loved the Cold War too much to let it go. By the end of the 1970s, Americans were completing a turbulent decade and a half that was the worst period most of them could remember. After the wrenchings and twistings of the 1960s and 1970s, Americans wanted off the roller coaster and onto solid ground. It wasn't coincidence that the 1970s witnessed a wave of nostalgia for the 1950s, that era of perceived stability on the other side of Vietnam and Watergate, when few Americans felt the need to question received wisdom regarding the world and their place in it. Nor was it coincidence that the era they looked on with such longing marked the time of the most thorough American embrace of the beliefs that formed the basis of the Cold War: that the United States headed the forces of good, that the Soviet Union chaired the party of evil, that communism must be resisted by all feasible means, that military strength was the best guarantee of American security, that compromise was appeasement and would lead to disaster. It was beside the point that the decade of Eisenhower had never been so placid as it was remembered to be, and that the predominant value system of the decade was as much a consequence of the events of the period as a cause. The salient fact in the late 1970s was that many Americans remembered the earlier era as a time of self-confidence and national well-being. In a present of persistent self-doubt and apparently increasing national ill-being, they were willing to listen to persons who promised a return to that past assurance, and who pledged to make America great again.

Ronald Reagan could have run for president, probably successfully, without even mentioning foreign affairs. Foreign affairs almost never decide American elections. The condition of the domestic economy usually does, for the reason that while some voters pay attention to whether and how the United States is involved in matters overseas, they all pay attention to whether and how they will be able to pay their bills. The unprecedented combination of inflation and unemployment of the late 1970s by itself would have sent Carter back to Georgia. The hostage crisis in Iran, the Soviet presence in Afghanistan, and sundry other manifestations of American ineffectualness abroad merely made Reagan's task easier.

Perhaps surprisingly, it didn't help Carter that by 1980 the president's position on foreign affairs, especially on the litmus issue of policy toward the Soviet Union, differed little from Reagan's. If anything, it may have hurt him. Reagan, in criticizing Carter's handling of international relations, naturally focused on the actions of the early Carter, the Carter of detente. Carter, having himself rejected detente, was poorly placed to refute the charges. By his rejection of his earlier views, he seemed in fact to confirm the criticism being brought against him.

Regardless of whether Americans who went to the polls in November 1980 voted principally against stagflation, or in favor of a return to the Cold War, they elected a president committed to the latter. As Reagan made plain throughout the campaign, and as he continued to make plain during his first months in office, he believed that a willingness to wage Cold War against the communists was a vital measure of America's capacity to reclaim the greatness of former days.

Yet the Cold War Reagan had in mind, or at least the Cold War he put into practice, was one with a crucial difference from what had gone previously. For all the psychological reassurance the Cold War of the period from the late 1940s through the late 1960s had afforded Americans, for all Keynesian impetus it had furnished the American economy, and for all the political distractions it had supplied supporters of the domestic status quo, the original Cold War had suffered from one serious disability. Namely, it repeatedly put large numbers of Americans—sometimes the entire population—at serious risk of their lives. Korea killed more than 50,000 Americans, Vietnam nearly 60,000. The several Berlin crises, the pair of face-offs in the Taiwan Strait, and the Cuban missile showdown pushed the superpowers alarmingly near to superwar. Whatever the advantages of the Cold War, this war-producing and war-risking tendency was a distinct disadvantage.

Reagan's Cold War represented a major improvement. While it preserved most of the psychological, economic, and political benefits of the first Cold War, it avoided most of the dangers. The remarkable thing about Reagan's initial term in office, when the Republican president denounced the Soviet "evil empire" with every second breath, and vowed unrelenting opposition to the Kremlin, was that the United States and the Soviet Union never came anywhere close to a direct confrontation. Even detente, which had witnessed the Defcon 3 alert of the Yom Kippur War, produced greater tension than the superpowers ever experienced under Reagan. For all his ideology-laced stump speeches, and his elevation of neoconservative fire-eaters to prominent posts in his administration, Reagan fought the Cold War primarily as an exercise in symbolism. His rhetoric suggested war, but his actions stayed well away from everything that carried a real danger of war.

Two events of Reagan's first term provided opportunities for going head to head with the Kremlin, and in each case the president settled for a symbolic thrashing of the communists. In December 1981, following months of unrest in Poland instigated by the opposition Solidarity movement, the Polish military staged what amounted to a coup, imposed martial law, banned Solidarity, and imprisoned its leaders. Though the Soviet hand was not so evident this time as it had been in Hungary in 1956 or Czechoslovakia in 1968, few Western observers doubted that General Jaruzelski and his Military Committee of National Salvation were doing Moscow's bidding, if only to forestall Moscow's doing the bidding itself. The Reagan administration responded with expressions of moral outrage, but its substantive reaction was decidedly mild—and was directed chiefly against hapless Poland, rather than against the Soviet Union. The administration terminated discussions designed to restructure Poland's Western debt, halted government-sponsored shipments of agricultural goods to Poland, barred Polish airliners from landing at American airports and Polish vessels from fishing in American waters, suspended Poland's most-favored-nation status, and prevented Poland from receiving American commercial credits or new loans from the International Monetary Fund. American sanctions against the Soviet Union consisted of a denial of landing privileges to Aeroflot, nonrenewal of a scientific exchange agreement, curtailment of certain purchasing practices by Soviet industry, new restrictions on the sale of industrial and high-technology items, and the delay of further grain deals. Washington did *not* stop current shipments of grain. It did *not* halt talks on intermediate-range nuclear forces then under way in Geneva, or talks on human rights and related issues in Helsinki. It did *not* cancel a scheduled meeting between the secretary of state, Alexander Haig, and the Soviet foreign minister, Andrei Gromyko. Moreover, within months, the administration was backing away from even its mild reproof of Moscow. It negotiated a new grain deal with the Soviet government, and after the European allies complained about the American strictures on the sale of high-technology items—some of which the Europeans wanted to include as part of a gas-pipeline deal with the Kremlin—Washington dropped this part of the sanctions package as well.

The second instance in which American outrage remained chiefly rhetorical followed the shooting down of a Korean airliner by a Soviet fighter plane in Soviet airspace in August 1983. Reagan and other administration officials characterized the downing as an act of cold-blooded barbarism, alleging that the Soviets had known that the straying aircraft was a civilian plane but had ruthlessly rocketed it anyway. Despite all its finger-pointing, the administration confined its substantive response essentially to another suspension of Aeroflot landing privileges.

During the first few years of the Reagan presidency, the gap between the administration's affinity for strong talk and its aversion to strong action appeared, to outsiders anyway, an ad hoc affair. But at the end of November 1984—after the president had been safely re-elected—Defense Secretary Caspar Weinberger provided some hooks on which to hang expectations regarding American policy. Weinberger outlined a set of six conditions he considered prerequisites to the commitment of American military forces to combat overseas. First, the troops should be committed only in situations "vital to our national interest or that of our allies." Second, American civilian officials must be sufficiently sure of the issues involved to allow American forces to fight "wholeheartedly and with the clear intention of winning." On this issue, Weinberger added, "If we are unwilling to commit the forces or resources necessary to achieve our objectives, we should not commit them at all." Third, the objectives, both political and military, must be "clearly defined," and American officials must be able to see "precisely how our forces can accomplish those clearly defined objectives." Fourth, the American government must constantly monitor the relationship between the conflict in question and the interests at stake, and determine whether securing the interests still necessitated the use of American force. "Conditions and objectives invariably change during the course of a conflict. When they do change, then so must our combat requirements. We must continuously keep as a beacon light before us the basic questions: 'Is this conflict in our national interest? Does our national interest require us to fight, to use force of arms?' " Fifth, before the president committed American troops to combat, there had to exist "some reasonable assurance we will have the support of the American people and their elected representatives in Congress." The memory of the last war had burned itself deep into the Pentagon's institutional memory. "We cannot fight a battle with the Congress at home while asking our troops to win a war overseas, or, as in the case of Vietnam, in effect asking our troops not to win, but just to be there." Sixth, all diplomatic and other alternatives must be exhausted before sending in the troops. "The commitment of U.S. forces to combat should be a last resort."[19]

Not everyone in the Reagan administration accepted Weinberger's formula. Secretary of State George Shultz favored diplomatic solutions to crises, as secretaries of state usually do, but he made the persuasive argument that diplomats do best when backed by armor and aircraft. Ten days after Weinberger's address, Shultz offered a rebuttal. "Americans have sometimes tended to think that power and diplomacy are two distinct alternatives," Shultz said. "This reflects a fundamental misunderstanding. The truth is: power and diplomacy must always go together, or we will accomplish very little in this world. Power must always be guided by purpose. At the same time, the hard reality is that diplomacy not backed by strength will always be ineffectual at best, dangerous at worst." To Weinberger's six points, Shultz opposed three. The use of American power was legitimate, he asserted, when, first, "it can help liberate a people or support the yearning for freedom"; second, "its aim is to bring peace or to support peaceful processes"; and third, "it is applied with the greatest efforts to avoid unnecessary casualties and with a conscience troubled by the pain unavoidably inflicted."[20]

Reagan sided with Weinberger generally, but not exclusively. In 1982, the president ordered American troops to join a United Nations peacekeeping force in Lebanon. The deployment to Beirut failed to meet several of Weinberger's criteria, the most telling deficiency being an inadequate definition of what the troops were supposed to do. The marines weren't intended to fight (which meant that Weinberger's criteria, narrowly if unrealistically speaking, didn't actually apply to them). But when American air and naval units got embroiled in the Lebanese civil war, the marines in Beirut became targets for terrorist attack. The attack came in October 1983, and 241 of the marines were killed.

Not coincidentally, within two days of the attack on the marine barracks in Beirut, and in the midst of a fresh round of questioning of the utility of American military force, Reagan ordered an American invasion of Grenada. The president had been doing the political spadework for the Grenada operation for some time. Concerned at efforts by Grenada's prime minister, Maurice Bishop, to improve relations with Cuba and the Soviet Union, Reagan warned against "the tightening grip of the totalitarian left in Grenada." When Cuba and the Soviet Union assisted Bishop's government in the construction of what Washington deemed a suspiciously long airport runway—one suitable for long-range military aircraft (or, as Bishop asserted, for long-range tourist airliners)—the president demanded, "Who is it for?" He answered his own question: "The rapid buildup of Grenada's military potential is unrelated to any conceivable threat. . . . The Soviet-Cuban militarization of Grenada, in short, can only be seen as power projection into the region."[21]

The bad situation got worse, from Washington's perspective, when persons apparently more radical than Bishop overthrew the prime minister. The new regime, headed by what Reagan called "a brutal gang of leftist thugs," instituted what the president described as a "barbaric shoot-to-kill curfew," and otherwise offended against democratic practices. Amid the uncertainty and confusion, the Reagan administration had little difficulty discerning a threat to the safety of Americans in Grenada. Washington worried especially about a group of several hundred students enrolled at a medical school in the capital, St. George's.[22]

Partly as a result of the administration's fear that the students might be harmed, or, what would be worse politically, that they would be taken hostage, creating an Iran-style standoff; partly from a desire to reverse the leftward trend of events in Grenada, thereby also serving notice to leftists in other countries, notably Nicaragua and El Salvador, that American patience had limits; and partly due to American frustration at the recent bombing of the marines in Beirut, to which administration officials hadn't determined how to reply or even where to send the message—from this combination of reasons, the president ordered the invasion of Grenada.

The 6000 American troops experienced only modest trouble, though more than they should have, righting the situation in Grenada. To the debatable degree that a Grenada tilting toward Cuba and the Soviet Union would have challenged American security interests in the Caribbean, the operation scored a success. The Pentagon awarded ribbons and medals by the bushel, and the many Americans hungry for a victory of any sort applauded the president's decisive action.

But the very enthusiasm of the response suggested that the operation was chiefly a matter of theater, and the mismatch between the contestants—the most powerful nation in the world versus a country that required a good atlas to locate—underscored the Reagan administration's cautious policy regarding the use of American military force.

The caution became more apparent still during the first several weeks of 1984. The rhetoric remained as staunch as ever. Speaking of the continuing American presence in Lebanon, George Shultz declared, in January, "It's important to show the world that we have resolve." A week later, Undersecretary of State Lawrence Eagleburger denounced a congressional proposal mandating the withdrawal of the marines from Lebanon, saying the administration was "morally bound" to oppose the bill. Eagleburger related messages from conservative governments in the Middle East urging the United States to maintain troops in Lebanon, and he asserted, "In a case like this, walking away from a difficult problem doesn't solve it; it only postpones the day of reckoning. A victory in Lebanon for the forces of radicalism and extremism will only embolden them." On February 2, an interviewer asked Reagan for a reaction to a call by the speaker of the House of Representatives, Thomas O'Neill, to bring the American marines home. "He may be ready to surrender," the president responded, "but I'm not." Within another week, however, Reagan announced a "redeployment" of the marines from Beirut to their ships offshore. Administration officials explained that "redeployment" wasn't the same thing as withdrawal. "I want to emphasize that the decision is not simply a removal-of-the-Marines decision," Shultz said. But such was precisely what it soon became.[23]

In the two other significant cases of the use of American military force during the Reagan years, the president took care to avoid placing American troops on foreign ground where they could get shot at or blown up. A 1986 air raid on Tripoli, undertaken in response to evidence linking Libya to the bombing of a West Berlin restaurant, in which an American soldier died, was a quick, in-and-out affair. Although American officials didn't say so, the raid amounted to an assassination attempt on Libyan leader

Muammar Qaddafi, in that it targeted his home, among other sites. The bombs missed Qaddafi, but killed a small girl he identified as his daughter. Evidence uncovered after the raid suggested that Syria, as much as Libya, had been behind the Berlin bombing.

In 1987, amid the final convulsions of the Iran-Iraq war, Reagan sent American warships to the Persian Gulf to protect the flow of friendly oil. This deployment produced few intentional casualties, although 37 American sailors died when a rocket fired by a plane from Iraq—America's de facto ally at the time—ripped into their ship, and 290 Iranian civilians were killed when an American vessel shot down an Iranian airliner by mistake.

Reagan's strategy of speaking loudly but keeping Americans mostly out of harm's way was shrewd. It provided the psychological satisfaction of blaming foreigners, particularly the Soviets and their allies and agents, for nearly everything wrong with the world—a vast improvement over Carter's search for demons within. It restored to the official American worldview the simplicity that had been lacking since the original Cold War consensus began unraveling in the 1960s. It afforded justification for a major campaign of military spending, at a time when many other government programs were taking sizable cuts. That an administration that constantly harped on the need for more and better weapons should be so reluctant to use them, and that the chief harper—Weinberger—should also be the most reluctant user, were simply two ironies among many of the Reagan years. Beyond the irony, these facts indicated that the defense buildup of the 1980s, like previous buildups, had purposes other than defense. Yet even as the anti-communist chest-beating restored American self-confidence and fueled a new round of military Keynesianism, the administration's caution allowed it and the United States to avoid the hazards of sustained violence. The arrangement was far more satisfactory than Carter's insipid introspection or Nixon's passionless but not hazardless geopolitics.

Reagan didn't disdain detente so thoroughly as to eschew everything associated with its author. In those cases where he believed that sustained violence was in order, Reagan followed Nixon's example in letting foreigners do the bleeding and dying. Further, like Nixon with regard to Cambodia and Chile (and Eisenhower with regard to Iran and Guatemala), Reagan had an affinity for acting covertly, so that the full degree of American complicity in controversial actions might remain hidden. By Reagan's turn in the White House, every full-term president since 1945 had had a doctrine named for him (although perhaps only aficionados—a clue here—could recall what the Johnson Doctrine was about). Notwithstanding his age and early thoughts of retiring after four years, Reagan now intended to go for eight, which made a Reagan doctrine the more mandatory. The one he, or rather his supporters, formulated posited the necessity for American backing of anti-communist guerrillas waging war on leftist governments. As the president got around to summarizing it in 1985, "We must not break faith with those who are risking their lives—on every continent, from Afghanistan to Nicaragua—to defy Soviet-supported aggression and secure rights which have been ours from birth." Employing his favorite euphemism for anti-communist rebels, he declared, "Support for freedom fighters is self-defense."[24]

The Reagan Doctrine might have been considered a corollary to the Nixon Doctrine, or perhaps an extension of the Truman Doctrine via the Doolittle lemma. Whatever its antecedents, the Reagan Doctrine was put to the test in four locales. In Angola, the United States once again supplied arms and other necessities of rebellion to the troops of Jonas Savimbi. In Afghanistan, American weapons enabled the anti-Soviet mujahideen to carry on their war against the Russian interlopers and their Afghan collaborators. In Cambodia, the United States provided aid to a coalition of resistance groups trying to root out the government installed by Vietnam following a 1978 invasion. In Nicaragua, the CIA oversaw the creation of the anti-Sandinista contra army, providing material and logistical assistance and helpful advice.

The Reagan Doctrine possessed some of the same advantages as the administration's tough-talk, soft-walk strategy toward the Soviets—of which it was, in fact, a subsidiary part. Americans could feel the warm glow of contributing to the fight against communism, without suffering the anguish of having to meet flag-draped coffins coming home. Because the wars America supported were confusingly irregular affairs, in which it was often impossible to know who was responsible for what, atrocities that occurred could be explained away. The cost of the wars to American taxpayers wasn't great, and even this cost wasn't fully visible, since the administration folded a sizable portion of the expense into the secret accounts of the CIA and other bureaus.

But in proxy conflict, as in everything else, you get what you pay for. In none of the four cases did America's guerrilla clients succeed in overthrowing the governments they targeted. An argument could be made, and certainly was, that without pressure from the contras, the Sandinista government of Nicaragua wouldn't have agreed to the 1989 election that led to the Sandinistas' peaceful ouster. At the same time, an argument could be made, and was, that without the the contra war, the Sandinistas would have held elections sooner, and wouldn't have engaged in all the repression they did. The Afghan mujahideen convinced the Soviets that their occupation of Afghanistan was unsustainable, but Mikhail Gorbachev came to a similar conclusion regarding Hungary and Poland and the rest of Eastern Europe even without the encouragement of American Stinger missiles. In any event, the Soviet departure left the mujahideen squabbling interminably among themselves. Jonas Savimbi failed to oust the Marxist government of Angola, and suspended his insurgency. A power-sharing arrangement in Cambodia, negotiated in late 1991, soon broke down.

Failure to achieve results wouldn't by itself have disqualified Reagan's Third World strategy. Some goals are difficult to reach. More troubling was the degree to which the Reagan administration staked America's reputation on groups that hardly represented what America professed to stand for. The problem was as old as the Cold War, of course. But by overblown remarks like his description of the Nicaraguan contras as modern equivalents of America's Founding Fathers, Reagan simply highlighted the distance between the noble ends America pursued and the ignoble means it employed to pursue them. Far from being George Washingtons and John Adamses, the contras often seemed more like the Hessian mercenaries who had fought on the side of the British. As even their supporters admitted, the contras would have put down their weapons and quit fighting if the United States had stopped paying their way. Jonas Savimbi and UNITA had a distressing habit of destroying American-owned petroleum facilities in Angola. In addition, Savimbi relied heavily on apartheid South Africa. After South Africa withdrew from the Angolan civil war, Savimbi called it a day. The mujahideen didn't even pretend to want to establish a liberal pluralist order in Afghanistan, which had never been so blessed. The resistance coalition the United States supported in Cambodia comprised some relatively benign, if opportunistic, characters like Norodom Sihanouk, but it also included the Khmer Rouge, the only outfit to challenge Hitler and Stalin for the murder championship of the century. (Adjusted for population size, the Red Cambodians won going away.)

Another weakness of the Reagan administration's approach to Third World affairs followed partly from the terms of the Reagan Doctrine and partly from the president's hands-off style of decision-making. By relying so heavily on covert programs, Reagan short-circuited the process of intra-administration, congressional, and public cross-checking that tends to keep American policy, if not always on the straight and narrow, at least from wandering too far into the wilderness of folly. And by delegating control of these covert programs to individuals of limited experience and less wisdom, the president exacerbated the problems arising from the lack of oversight. Reagan's inclination to covert action, and his disinclination to close scrutiny of subordinates, came together most damagingly in the Iran-contra fiasco, which bespattered his presidency and eroded American credibility more than any other event of his eight years.

But Reagan had mastered the art of lowering expectations. What would have destroyed the reputation of one of his predecessors—presidents whom the American people held to a reasonable standard of accountability— glanced harmlessly off Reagan. Truman had boasted that the buck stopped in the Oval Office. Eisenhower owned up to ordering Francis Gary Powers's U-2 flight. Kennedy accepted blame for the Bay of Pigs. Nixon lied about Watergate, but didn't get away with it. Whether Reagan was lying about his role in selling arms to Iran and diverting the profits to the contras, or really didn't know what his administration was doing, didn't appear to matter. Americans liked the old actor, and wanted to keep liking him.

The business of America

What *did* matter, as always, was the condition of the American economy. After an inflation-wringing recession in late 1981 and 1982 (largely the work of the tight-money policies of Carter-appointee Paul Volcker and the Federal Reserve), the economy entered a prolonged period of expansion. The country's gross national product climbed, in constant 1982 dollars, from $3.17 trillion in 1982 to $3.50 trillion in 1984, to $3.72 trillion in 1986, to $4.02 trillion in 1988. Unemployment declined every year of Reagan's presidency after 1983, from a high of 9.5 percent to 5.4 percent in 1988. Average personal income rose from $9,725 in 1982 to $11,337 in 1988 (in constant 1982 dollars), although much of the rise resulted from the greater participation of women in the paid labor force. Nearly 57 percent of women 16 years and older worked for pay in 1988, as opposed to 51.5 percent in 1980. Inflation eased markedly, with the annual increase in the consumer price index dropping from 13.5 percent in 1980 to a low of 1.9 percent in 1986, before climbing back to 4.1 percent in 1988.[25]

The president and administration true-believers attributed the economic good times to the astuteness of their conservative ideas. Reagan entered the White House committed not only to the restoration of a previous generation's views on international affairs but to a restoration of former relationships between the American government and the American economy. The two aspects of the Reagan restoration fitted neatly together. Indeed, they were inseparable. To the Reaganite mind, the enemy was the state. On the world scene, the enemy took the form of communist states, which carried the idea of statism to a totalitarian extreme. On the domestic front, the enemy adopted the guise of the welfare state, which, though not yet the Big Brother of communism, had been tending in that direction since the New Deal. Just as American foreign policy had the goal of containing and diminishing the power of communist states, so American domestic policy must have the goal of containing and diminishing the power of the American welfare state. In each case, the goal was to free the energies of individuals, releasing them from the oppressive weight of government control.

What Reagan's Cold War was to American foreign policy, Reaganomics was to American domestic policy. The central feature of Reaganomics was a passion for cutting tax rates. From the indisputable premise that confiscatory tax schedules decreased individuals' desire to work, the administration proceeded, by way of the Laffer curve, to the contestable conclusion that cutting taxes would increase both economic activity and government revenues. (What was missing in the chain of deduction was concrete evidence that current tax rates were confiscatory in the Laffer sense. Some administration officials found this missing link troubling, as budget director David Stockman later admitted.) The extra revenue would be supplemented by cuts in non-defense spending—such cuts being an important part of the campaign to get the government off the backs of the people. Economic deregulation, another minimal-government imperative, would similarly stimulate the economy, providing greater prosperity and, as a side effect, still more government revenue. The administration would use the revenues so generated for one of the few purposes the Reaganites conceded to be within the legitimate scope of government: national defense. Without having to increase the burden on individuals, the administration would restore America's military might.

In practice, things didn't work out quite as the administration planned. It got the cuts it wanted in income taxes in 1981, but subsequent increases in social security taxes offset most or all of the cuts for many persons of modest means. It got much of the deregulation it aimed for, but the cost of deregulation eventually included a monster bail-out of the savings and loan industry. Congress accepted some of the non-defense budget trimming, but the legislature drew the line well short of where the president thought the line ought to be. The stimulated economy generated more revenues than before, but the revenues didn't measure up to expenditures, and the government ran unprecedented deficits.

Certain groups could hardly complain. The restructuring of the country's tax system—in particular the large shift of collections from the progressive income tax to the regressive social security tax—gave a windfall to the rich. Between 1977 (the year of an earlier boost in social security taxes) and 1984, the average federal tax rate, covering federal taxes of all kinds, paid by the poorest 10 percent of Americans *increased* by more than 30 percent, while the rate paid by the wealthiest 10 percent *decreased* by nearly 12 percent. The extremely rich did even better, with the rate on the wealthiest 5 percent falling by more than 15 percent and the rate on the wealthiest 1 percent falling by more than 25 percent. At the same time, the huge growth in the national debt, and consequently in the interest paid to service the debt, implied an additional large redistribution of wealth from Americans who didn't own government securities to those who did—in other words, from the less affluent to the more affluent. The great growth in the debt tended to push all real interest rates up, adding again to the income of the rentier class. One imperfect (in light of inflation), but not inconsiderable, measure of the improving fortunes of the wealthy during the Reagan years was the millionaire index, which registered approximately 574,000 millionaires in the United States in 1980, 832,000 in 1985, and 1,500,000 in 1988. (Once more, the *really* rich did the best: the number of American billionaires increased from less than 5 in 1980 to more than 50 in 1988.)[26]

Needless to say, the boom times for the well-heeled resulted from circumstances that went far beyond Reagan's revival of the Cold War. Yet there were some obvious connections, for example between increased military expenditures and the rises in the federal debt and interest rates that benefited the coupon-clippers. More generally, and more importantly, the complex of values Reagan represented, both domestic and foreign, fostered conditions in which the wealthy thrived. This didn't necessarily make the wealthy into raving Cold Warriors, but it did make them think twice about tampering with a system that was serving them so well.

For one segment of the economy, the connection between the revived Cold War and prosperity was direct and unmistakable. During the first half of the 1980s, American spending on defense went up rapidly, with outlays jumping by 34 percent in real terms between fiscal 1981 and 1985, from $171 billion to $229 billion (1982 dollars). After the lean years of the post-Vietnam, detente era, the Pentagon found itself gorged with funds. The admirals laid plans and keels for a 600-ship navy. The air generals ordered bombers, advanced fighters, and cruise missiles by the score and hundred. The army generals took delivery of trainloads of faster and more powerful tanks, computer-guided rockets, and state-of-the-art equipment of all sorts.[27]

Whatever it accomplished for American defense, the Reagan buildup worked wonders for the American defense industry. Between the late 1960s and the late 1970s, companies that did business with the Pentagon had been in a major collective slump. Stock prices of Standard & Poor's aerospace/defense sector (which included such household names as Boeing, General Dynamics, Lockheed, Martin-Marietta, McDonnell Douglas, Rockwell, and United Technologies) fell by a third during this period, with prices at certain points being off by nearly three-fourths. Profit margins showed a similar decline. Jimmy Carter's late conversion to anticommunist confrontation sent prices sharply upward, and the advent of the Republican administration extended the trend. In a generally bullish time, prices for aerospace/defense issues outpaced the stock market as a whole by a sizable amount. Average profit margins rose from 6.8 percent in 1975 to 9.6 percent in 1985. Dollar earnings per share, after stumbling along in the middle and low single digits through the mid-1970s, spent the Reagan years almost exclusively in the 20s and 30s.[28]

Nor was the new Cold War good business only for those directly involved in making the weapons of war. During the Reagan years, federal expenditures on defense-related research and development more than doubled, from $18.4 billion in 1981 to $39.2 billion in 1987. Though much of this money went to business corporations, a substantial portion found its way to individuals associated with leading universities. Vietnam-era sensitivities about classified research on campus had led some universities to push Pentagon projects to affiliates like Lincoln Laboratories at MIT and SRI International at Stanford, but enterprising professors had little difficulty working out consulting arrangements. Meanwhile, straightforward research payments by the defense department to universities increased by nearly 90 percent between 1980 and 1985.[29]

The constituency for a big defense establishment increased still more in March 1983, when Reagan announced plans for developing a defense against strategic nuclear weapons. Describing the policy of deterrence—peace on pain of nuclear annihilation—as "a sad commentary on the human condition," Reagan asked, "Would it not be better to save lives than to avenge them? Are we not capable of demonstrating our peaceful intentions by applying all our abilities and our ingenuity to achieving a truly lasting stability?" Replying to his own question, he said, "I think we are—indeed, we must."[30]

The star wars gambit—officially, the Strategic Defense Initiative—was a brilliant maneuver, for it cast opponents as hidebound Dr. Strangeloves who would consign humanity to the never-ending threat of imminent destruction. Nonetheless, opposition soon arose. The most distinguished group of critics called itself the Union of Concerned Scientists. In October 1984, UCS members Hans Bethe (a Nobel-prize physicist), Richard Garwin, Kurt Gottfried, and Henry Kendall published a detailed critique of SDI in *Scientific American*. The four pointed out that a defense like that which the president envisioned would have to be able to intercept, almost simultaneously, virtually all the 10,000 or so warheads the Soviet Union might commit to a major attack. And it would have to accomplish this remarkable feat the first time, without any tests approaching real conditions. The scientist-authors reviewed the feasibility of various methods proposed by star-wars advocates for shooting down Soviet missiles during the crucial boost-phase, when the warheads remained attached to their easily spotted rocket boosters, and before multiple warheads diverged on separate descents. They assessed the likelihood of improvements in technologies relating to lasers, particle beams, and X-rays, to the necessary level of sophistication. They estimated the costs of deployment, and the costs of Soviet countermeasures such as decoys, missile-shielding, altered trajectories, and construction of more offensive missiles. They expressed concern that a star-wars race would produce dangerously effective anti-satellite weapons. And they concluded that while SDI wasn't intrinsically unachievable, for the foreseeable future it was practically so. "In our view the questionable performance of the proposed defense, the ease with which it could be overwhelmed or circumvented and its potential as an antisatellite system would cause grievous damage to the security of the U.S. if the Strategic Defense Initiative were to be pursued."[31]

Star-wars proponents counterattacked. They trotted out their own experts, including Dartmouth astrophysicist Robert Jastrow, who found a predictably friendly vehicle for his views in *Commentary*. As often happened aboard that neoconservative flagship, Jastrow not only impeached his opponents' testimony but impugned their motives as well. He asserted that the UCS had got crucial figures on SDI, for instance involving the number of satellites required to repel a Soviet strike, all wrong. Other SDI critics had made comparable mistakes, which Jastrow corrected. "How did published work by competent scientists come to have so many major errors?" he asked. And why did the errors all lead in the same direction— "toward a bigger and more costly defense, and a negative verdict on the soundness of a 'Star Wars' defense against Soviet missiles"? Jastrow was willing to defer to the likes of Bethe on matters within Bethe's realm of expertise, but on SDI, he said, Bethe was out of his depth. Jastrow approvingly quoted a colleague in weapons analysis: "Is Hans Bethe a good physicist? Yes, he's one of the best alive. Is he a rocket engineer? No. Is he a military-systems engineer? No. Is he a general? No. Everybody around here respects Hans Bethe enormously as a physicist. But weapons are my profession. He dabbles as a military-systems analyst." To some degree, Jastrow said, the bias in the errors of the SDI critics reflected this kind of dilettantism. Yet it also demonstrated that scientists ground axes just like everyone else. "As with the rest of us, in matters on which they have strong feelings, their rational judgments can be clouded by their ideological preconceptions." These ideological preconceptions caused the critics to accept errors that an impartial observer could easily detect. "I would like to wager that if the theorists studying the matter for the UCS had found that only 10 satellites could protect the United States from a massive Soviet attack—if they had gotten a result that indicated that the President's proposal was simple, effective, and inexpensive to carry out—then they would have scrutinized the calculations very, very carefully."[32]

With the experts split, non-experts often felt free to take whichever side of the debate suited their interests. For many people, the idea of pulling the world out from under the nuclear cloud held enormous appeal. Others—as well as many of the same people—responded to considerations less high-minded. What critics cited as a principal defect in SDI—its high cost—in fact proved to be a major generator of support. Backers of the program, beyond claiming its strategic necessity, pointed to the benefits it would confer upon the economy as a whole. General James Abrahamson, chief cheerleader of the enthusiastically pro-star-wars Strategic Defense Initiative Organization, asserted, "Relative to SDI, computer, communications, propulsion, and laser technologies have attractive and significant spinoff possibilities." In a variant of supply-side economics, Abrahamson added, "Clearly they would help the SDI program pay for itself." Congressman Kenneth Kramer, a Republican from Colorado, proclaimed that SDI and associated activities would "lay the foundation for an educational-vocational renaissance for the American labor force, particularly the unemployed in the 'smokestack industries.' " Kramer continued, "When we implement the defense initiative, it becomes the largest single undertaking ever attempted by mankind." He proudly predicted that Colorado Springs, his home and the site of the new Unified Space Command, would soon be "the space capital of the free world."[33]

Once the Reagan administration demonstrated a firm intent to pursue the star-wars concept, applications for a piece of the action began to pour in. By the summer of 1985, the administration's SDI coordinating office had received over 1000 proposals for research grants. Most came from university scientists. "Virtually everyone, on every campus, wants to get involved," James Ionson, the official in charge of setting up research consortiums relating to SDI, told a reporter, with undisguised pleasure. "There will be many, many Manhattan projects in this."[34]

Whether as a result of an even distribution of research talent in American universities, or, more likely, from a desire to spread the wealth into as many congressional districts as possible, the Pentagon channeled SDI funding to campuses all across the country. Between 1983 and 1986, Utah's higher education system led the way in SDI contracts, with more than $36 million, followed closely by the University of Texas, Georgia Tech, and Stanford, each of which received more than $20 million. Sixteen other universities or university systems, in eleven states, also topped the one-million-dollar mark. Within the next two years, the number of campuses getting SDI funding passed eighty.[35]

The research laboratories that shared personnel with the universities did even better—very much better. The Lawrence Livermore and Los Alamos laboratories, administered by the University of California, between them garnered 39 SDI contracts during 1983–86, worth more than $1 billion. MIT's Lincoln Lab pulled down $327 million, while Caltech's Jet Propulsion Lab and the University of Chicago's Argonne Lab received smaller but still substantial pieces of the pie.[36]

Weapons contractors did best of all. The *Wall Street Journal* summarized the corporate reaction to the president's proposal for strategic defense: "For defense contractors across America, President Reagan's Star Wars program is more than a new strategy for defense. It is the business opportunity of a generation, a chance to cash in on billions of dollars in federal contracts." The *Journal* went on to cite a defense-industry expert who described the rush for contracts as "a fish-feeding frenzy." An executive with a defense firm told the *Christian Science Monitor* as early as 1984, "All of the major aerospace industries are deeply and heavily into this now, and all of them view this as an opportunity that they can't pass up. Twenty-five billion dollars is an awful lot of money. That's a strong incentive for any company to pursue it, even if Congress cuts it in half and we only get 10 percent." A defense analyst for First Boston Corporation explained the anxiety to get in on the ground floor of SDI as a function of a hunch that the traditional defense budget wouldn't continue to grow forever. "Every company is on notice that, if they want to be a long-term player, they can't let SDI get away."[37]

They didn't. Through 1986, Lockheed received $717 million in SDI contracts. General Motors, which saw a good thing coming and snatched up Hughes Aircraft, got $547 million. Boeing, TRW, McDonnell Douglas, General Electric, and Rockwell each hauled in more than $200 million. Altogether, thirty-six corporations received more than $20 million apiece in SDI contracts.[38]

The officials responsible for dishing out SDI monies continued to spread the wealth widely. Forty of the 50 states received funding from SDI programs. While 43 states suffered net losses from SDI—meaning that they got less back in contracts than they paid as SDI's portion of their taxes—it was in the nature of American politics that the beneficiaries felt the benefits, in terms of jobs and profits, more than the injured felt the injuries, in terms of higher taxes.

It was also in the nature of American politics that political action committees associated with leading defense contractors contributed lavishly to the campaigns of legislators in positions to decide who got how many of the SDI dollars. In the 1983–84 election cycle, the 57 representatives on the House of Representatives' armed services committee or the defense subcommittee of the appropriations committee—comprising just 13 percent of House membership—received 35 percent of total SDI-contractor PAC contributions. In the Senate, the disproportion was less but still significant. There the 36 members of the relevant committees—36 percent of Senate membership—received 56 percent of star-wars PAC money.

These favored legislators returned the favors, particularly to defense contractors in their home states and districts. Even while broadcasting SDI appropriations over all but 10 states, the key committees concentrated funding in members' states and districts. In 1983–84, for example, states with senators on the armed services committee or the appropriations defense subcommittee received 87 percent of prime contracts for SDI work. In 1985–86, the figure rose to 92 percent.[39]

To skeptics, the SDI boom showed a familiar pattern. Paul Warnke, the chief arms-control negotiator in the Carter administration, asserted, "What we see happening today is the rapid conversion of the President's Star Wars proposal from stardust and moonbeams to that great pork barrel in the sky." George Ball, the former undersecretary of state, explained, "Perhaps the most effective support for Star Wars is being generated not by ideology but by good, free-enterprise greed. Firms in the hypertrophic defense industry, along with their thousands of technicians, are manifesting a deep, patriotic enthusiasm for Star Wars. Since they are experienced in lobbying and wield heavy influence with members of Congress who have defense plants in their constituencies, they are creating a formidable momentum for the project. Whether or not it would contribute to the security of the nation, it offers *them* security."[40]

Right revisionism

Partly as a consequence of the military buildup, partly as a result of the tax cuts, and partly from the refusal of Congress to trim non-defense spending as much as the president desired, the Reagan administration ushered in a new era of American federal financing. The hallmark of the era was a large and seemingly permanent budget deficit. The deficit first hit three figures (in billions of dollars) in fiscal 1982, and after that it stayed above $150 billion, going as high as $221 billion in 1986 (and much higher after Reagan left office). As a result, interest on the national debt composed the fastest growing portion of the federal budget, rising by more than half between 1980 and 1988, despite a sizable decline in interest rates.[41]

Although officials of the Reagan administration derided the allegedly liberal theories of John Maynard Keynes, the coincidence of the massive federal deficits with the economic expansion of the 1980s was certainly consistent with Keynes's demand-side arguments. Many Republicans, after decades of describing Democratic deficits as a failure of government, had to revise their views in light of their own party's unprecedented contribution to the nation's outstanding debt. Reagan stuck to his anti-government guns, blaming the deficit on Congress, but other conservatives began contending that the deficit didn't really matter as much as people had once thought. Besides, in an economy as big as America's, big numbers could be misleading. The important figure, they said, was the size of the deficit relative to the economy as a whole. This had increased only modestly over the previous quarter-century, from 19.2 percent in 1962 to 22.8 percent in 1987. Still other analysts asserted that American accounting practices muddled the figures. State and local surpluses, they said, largely offset the federal deficit. Moreover, the federal budget didn't distinguish between investment, as on roads and education, which would return dividends over time, and consumption, which wouldn't.[42]

Even as they challenged received wisdom regarding the deficit, Reagan's partisans revised the history of the Cold War. While Reaganite economics absolved the current generation of responsibility for paying for the new Cold War, Reaganite—or neoconservative—history absolved it and its predecessors of responsibility for the mistakes of the old Cold War. Not all the rewriting was neoconservative. Some historians and political scientists merely corrected the extreme claims of the New Leftists of the 1960s and early 1970s, and their work represented a search for middle ground between the blame-the-Russians approach of the early accounts of the Cold War and the blame-the-Americans approach of the New Left.

The neoconservatives, by contrast, rejected the middle ground. They did so especially on the most contentious issue in the American history of the Cold War: Vietnam. Predictably, Norman Podhoretz led the neoconservative counterattack. In a polemical—even by Podhoretzian standards—volume entitled *Why We Were in Vietnam*, the *Commentary* editor essayed to do for Vietnam what loyal Southerners had long done for the American Civil War: to portray it as a failed but nonetheless honorable endeavor. Podhoretz rejected the charge that the initial commitment to Vietnam was based on a misconception. "In subsequent years," he wrote, "much ridicule came to be heaped upon the idea of an international Communist movement or a monolithic Communist conspiracy, but there was nothing ridiculous or even overstated about this idea in 1949." Impatient at the liberals' late conversion to dovedom, Podhoretz delighted in demonstrating that the war in Vietnam had started with Kennedy. "Kennedy and his people took it as virtually self-evident that the lesson of Munich was as applicable to the war in Vietnam as it had been to the war in Korea"—another Democratic conflict. "The United States, under Kennedy, for all practical purposes and in all but name, went to war in Vietnam."

Podhoretz spent little time on what was, for his purposes, the technical issue of the war's winnability. This he would leave to the growing phalanx of military second-guessers. He was far more interested in questions of morality, especially the fundamental question of whether American intervention in Vietnam had been *right*. In answering this question—affirmatively— Podhoretz tipped his hand regarding the underlying purpose of his polemic. To neoconservatives, the rehabilitation of the Vietnam experience was principally a device for reaffirming the ideological premises of the Cold War. Why had the United States gone into Vietnam? Because not doing so would have conceded all of that country to the dictatorial whims of Ho Chi Minh. Why did the United States stay in Vietnam so long? Because the original purpose was just, because the United States couldn't lightly abandon its Vietnamese allies, and because pulling out risked throwing the Free World into confusion and despair.

These were the kinds of questions that had been—and would be— posed every time the American government considered intervention in foreign quarrels. Though Podhoretz couldn't deny that Vietnam had turned out badly, he sought to repair the damage Vietnam had done to Americans' willingness to confront communism. He branded most critics of the war as "anti-American," defining the term to suit his purposes. "By anti-Americanism I mean the idea that the United States was a force for evil in Vietnam." He blasted Jimmy Carter for describing the Vietnam War as signifying the "intellectual and moral poverty" of American foreign policy. He embraced the view of Ronald Reagan—no surprise here—that Vietnam was a "noble cause." The worst that could be said of America's effort in Vietnam, Podhoretz declared, was that it was an act of "imprudent idealism." The United States might exercise greater prudence in the future, but it mustn't surrender its idealism to the naysayers and the appeasers.[43]

Other neoconservatives joined Podhoretz in combatting what they generically called "the new isolationism." Charles Krauthammer, a regular contributor to the increasingly neoconservative (especially in foreign affairs) *New Republic*, offered a modest taxonomy of this pernicious philosophy. "Left-isolationism," originally an affliction of McGovernites, but recently found throughout the party of Jefferson and Jackson (Old Hickory must have been cursing in his grave) cloaked its true isolationist nature behind a veil of internationalism. The left-isolationists, Krauthammer stated, acknowledged that America had something to offer the world, but they didn't believe that the United States should force its values and institutions on other countries. After quoting Democratic senator Daniel Patrick Moynihan, who had criticized the American invasion of Grenada on the reasoning that "you don't bring in democracy at the point of a bayonet," Krauthammer responded, "That idea will come as a surprise to Germans and Japanese." At those Democrats who asserted the need for a multilateral approach to world problems, and pointed to the successful multilateralism of the Korean War, Krauthammer scoffed. "For the strong, multilateralism is a cover for unilateralism. For the rest it is a cover for inaction. Today the United States is part of the rest."

Krauthammer's "right-isolationists" were harder to locate in terms of party. They included the arch-detentist Henry Kissinger, who had spoken of reducing the number of American troops in Europe; Irving Kristol, the editor of *Public Interest* (and a committed neoconservative on most issues), who agreed with Kissinger on this matter; the Reagan administration's own Caspar Weinberger, with his paralyzing restrictions on the use of American military force; and a variety of others. For Krauthammer, Weinberger's position summarized the attitude of the right-isolationists generally. They didn't reject the use of force, in principle, but they boxed so tightly the circumstances under which they would use force—Weinberger's notorious six conditions being the clearest case of this boxing—that they might as well have.

Krauthammer granted the allure of isolationism, whether of left or right. But no one had ever said the path of virtue would be easy. For Americans to be true to themselves and to their national purpose, he explained, they must close their ears to those who would define American interests in strictly nationalistic terms, and would leave the world to its fate. "To disengage in the service of a narrow nationalism is a fine foreign policy for a minor regional power, which the United States once was and which, say, Canada and Sweden are now. For America today it is a betrayal of its idea of itself."[44]

George Will was not a neoconservative, nor a neo-anything-else. Will preferred the company of Edmund Burke. Yet Will's books and syndicated columns complemented the neoconservatives' efforts on behalf of Reaganism. He had been a fervent opponent of SALT II, and he spent the first half of the 1980s warning against backsliding in arms-control's direction. He said that during the period of the SALT process, the American nuclear arsenal had stagnated, while the Soviet arsenal had grown "quantitatively and qualitatively." Noting the thousands of weapons the Soviets had added to their stocks in recent years, Will asked, "Does anyone think the world is safer than it was when the SALT process began?" He argued that arms control inevitably favored a closed society such as the Soviet Union, which could cheat undetected, while hurting an open society such as the United States, which couldn't. As long as negotiations continued, the Soviets conducted their affairs with accustomed ruthlessness. "The arms control era has coincided with unparalleled Soviet aggression and

threats, from Indochina through Afghanistan. Try to tell victims of 'yel-
low rain' [biological or chemical agents allegedly used by Soviet clients]
about the wonders of arms control. Biological weapons are controlled—on
paper. What has violation of the controls produced? A U.S. clamor for
more agreements." Such clamor, well intentioned or not, worked against
American interests by increasing the pressure for agreements, which trans-
lated into pressure for American concessions. "An immoderate and
unempirical belief in arms control," Will concluded, "produces a policy of
apologetic retreats."[45]

Praise the Lord and pass the ammunition

Podhoretz, Krauthammer, and Will occupied positions on the intellectual
wing of the conservative movement of the 1980s. Equally important—in
fact, more important in terms of sheer voting numbers—was the movement's non-intellectual (tending to anti–intellectual) wing. While the
neoconservative complaint against the politics and values of the 1960s and
1970s was essentially ideological, the animus of the New Right (a term
often used, but not always consistently, to distinguish the non-intellectuals
from the neoconservatives and fellow-travelers like Will) was primarily
cultural. In the thinking of the New Right, the principal danger to the
American way of life was something identified as secular humanism. A
group called the Pro-Family Forum compiled a list of what it said the
secular humanists stood for. Secular humanism, the list asserted:

> Denies the deity of God, the inspiration of the Bible and the divinity of Jesus
> Christ.
>   Denies the existence of the soul, life after death, salvation and heaven,
> damnation and hell.
>   Believes that there are no absolutes, no right, no wrong—that moral values
> are self-determined and situational. Do your own thing, "as long as it does
> not harm anyone else."
>   Believes in the removal of distinctive roles of male and female.
>   Believes in sexual freedom between consenting individuals, regardless of
> age, including premarital sex, homosexuality, lesbianism and incest.
>   Believes in the right to abortion, euthanasia (mercy killing) and suicide.
>   Believes in equal distribution of America's wealth to reduce poverty and
> bring about equality.
>   Believes in control of the environment, control of energy and its limitation.
>   Believes in removal of American patriotism and the free enterprise system,
> disarmament, and the creation of a one-world socialistic government.[46]

Whether or not anyone actually qualified as a secular humanist by all
these standards, the idea of secular humanism was very significant. A
politically potent number of Americans worried that the secular humanists
were trying to take over the country—and not just the country, but the
world, as the linking of atheism, situational ethics, homosexuality, abor-

tion, income redistribution, and environmentalism to disarmament and international socialism indicated. The opponents of secular humanism considered their entire constellation of values to be in danger. Not unnaturally, they tended to believe that giving ground on any one issue would weaken their position on other issues. In this way, correct thinking on the Bible and the Equal Rights Amendment became inextricably bound up with correct thinking on SALT, the Reagan Doctrine, and the Cold War generally.

While some spokespersons for the New Right contented themselves with attacking communism in unspecific language, others described their views on the Cold War in detail. For the most part, these views coincided with those of the neoconservatives and the Reagan administration. Jerry Falwell of the Moral Majority, writing in 1986, tallied up some of the high points of his organization's activities. "We have defended the Strategic Defense Initiative," Falwell said. "We have opposed an immediate unverifiable nuclear freeze. . . . We have supported financial aid for the Freedom Fighters in Nicaragua. We have openly opposed possible Communist takeovers in Taiwan, South Korea, the Philippines, South Africa and all over the world."[47]

Richard Viguerie, the most politically adept of the New Rightists—certainly the most adept at raising funds—delineated in greater detail the movement's agenda on foreign policy in a manifesto entitled *The New Right: We're Ready to Lead*. (Jerry Falwell wrote an introduction to the book, applauding Viguerie's "courage to speak out regarding liberals and their actions that have significantly occasioned America's perilous condition.") Viguerie explained that there existed a straightforward connection between the domestic aims of the New Right and its foreign policy goals: specifically, the conviction that unless Americans prevented communism from dominating the world, their concerns regarding domestic affairs would be for nothing. Describing various groups advocating the reversal of America's liberal deterioration, Viguerie wrote, "All these people realize that it's idle to talk about their goals and ideals unless America's survival is assured. And they agree that our survival has become a real question." Viguerie recited the litany of America's military decline vis-à-vis the Soviet Union. The Soviets had five times as many tanks, almost twice as many strategic offensive weapons (he didn't specify his source for this dubious statistic), and twice as many "major ships and submarines" (nor for this). He concluded, "Clearly, we have fallen from being the Number One military power in the world to the Number Two power, behind a country whose leaders are totally committed to defeating America and conquering the world."

Viguerie proposed several measures to prevent the Soviets from achieving their goal. First, the United States must "regain strategic military superiority without delay." Second, the American government must ban the transfer of goods, technology, and capital that might contribute to the

Soviet Union's capacity to wage war. Third, American officials should "expose and condemn" Soviet violations of international agreements and laws. Fourth, the American military must develop the conventional capabilities necessary to stop "Soviet-sponsored aggression." Fifth, Washington should employ diplomatic devices "like recalling our ambassador and severing diplomatic relations" to protest especially atrocious Soviet acts, such as the 1968 invasion of Czechoslovakia and the 1979 invasion of Afghanistan. Finally, the American government should take steps to "free America from arms control restraints which perpetuate U.S. military inferiority and force us to fight the Third World War by Soviet rules."

Viguerie went on to endorse a set of proposals recently forwarded by William Buckley's *National Review* and designed to liberate the "captive nations." For starters, the American representative at the United Nations should offer a resolution to the General Assembly demanding the withdrawal of Soviet troops from Eastern Europe. Next, the White House should announce that any future negotiations with Moscow would include the demand for the de-occupation of Poland, Hungary, and the other satellite countries. The United States should also call for the withdrawal of Russian troops from the Baltic republics and the non-Russian republics within the Soviet Union, and for free elections there and in Eastern Europe. Simultaneously, the American government should intensify its anticommunist propaganda activities, paying special attention to Radio Free Europe and Radio Liberty. "There is no good reason why we should not spend as much time and effort stirring up trouble behind the Iron Curtain as the Soviets spend trying to undermine our domestic system."

The heart of the problem, Viguerie said, was that most Americans didn't realize that there was a war going on, while the Soviets assuredly did. Americans must wake up to the danger. The FBI must increase its surveillance of subversives in the United States, just as it had during World War II. Viguerie cited as particularly telling a statistic (which, context suggested, came from the *Conservative Digest*) revealing a 99 percent drop in domestic security cases during the 1970s. "Meanwhile 1,900 registered Soviet bloc officials went about their spy business." War-style mobilization was imperative. "We must return to a World War II emergency method of rebuilding our military position. That would mean doing away with time-consuming purchasing requirements, arms control impact statements, low bids, and any EPA or OSHA rules that hampered production. It would mean keeping some plants open 24 hours a day. It would mean a total commitment by everyone from the President to the worker on the job and the realization that we are at war with a dangerous enemy." There was no room for choice. "The alternative to such an all-out American effort is simple. The Soviets will either force us into a war we will lose, or we will be forced to surrender."[48]

No one on the New Right wanted to surrender, but some weren't convinced a major war would be such a bad thing. Whether American millenarianism became more popular during the late 1970s and 1980s, or just more visible, is difficult to tell. But someone was buying the 18 million copies that Hal *Lindsey's Late Great Planet Earth* sold. Lindsey's thesis was that the end-time was nearing fast. His first clue was the regathering of Zion in Palestine, as the prophet Ezekiel had foretold 2600 years before. Lindsey adduced various bits of evidence to show that the creation of Israel in 1948 was what Ezekiel had been talking about, and he went on to say that the restoration of the Jews to their homeland had prepared the stage for the portentous events that would follow. "This physical restoration to the land is directly associated with triggering the hostility which brings about a great judgment upon all nations and the Messiah's return to set up God's Kingdom."

Israel's establishment had definitely triggered hostility, but Lindsey wasn't especially concerned with what the Palestinians and the other Arabs thought about the Zionists. The enemy that really counted was the one from "the uttermost north." In case readers' geography was weak, Lindsey explained, "There is only one nation to the 'uttermost north' of Israel—the U.S.S.R." Etymological evidence clinched Lindsey's case. Ezekiel had identified the powerful and malign Gog of the north as the ruler of Rosh, Meshech, and Tubal. The name Rosh, Lindsey asserted, on the authority of a chain of ancient and modern scholars, was none other than an archaic version of the word Russia. Meshech was the forerunner of Moscow. (Tubal didn't translate so neatly.) Lindsey thought it particularly noteworthy that certain Israeli leaders believed that their country's real foes were the Soviets. He quoted Moshe Dayan as saying, "The next war will not be with the Arabs but with the Russians." (Dayan held special significance for Lindsey and the millennialists, as the liberator of Jerusalem during the June War of 1967. According to prophecy, the recapture of the site of Solomon's temple brought the final days that much closer.)

Lindsey commented on other parallels between recent events and biblical prophecies. "The Bible says that Egypt, the Arabic nations, and countries of black Africa will form an alliance, a sphere of power which will be called the King of the South. Allied with Russia, the King of the North, this formidable confederacy will rise up against the restored state of Israel." Making his case prior to the Camp David accords (which apparently did little to change his mind), Lindsey asserted that the present lineup in the Middle East and Africa was pretty much what the prophet had ordered. Lindsey said that the book of Revelation predicted the emergence of a mighty nation in the east with an army of "200 million." This could only refer to Communist China, he declared. A fourth kingdom would arise in the west, a kingdom described as a new Rome. Number four, Lindsey explained, was the European Common Market, or, rather, the integrated community the Common Market was developing into. On this point, he quoted Dean Rusk, who when secretary of state had contended that the Common Market would become far more than an economic organization. "Powerful forces are moving in the European community toward political integration as well," Rusk said. "Survival and growth force the nations of Europe to forget their historic antagonisms and unite. Through the pooling of resources and efforts a mighty new entity is growing out of the chaos left by national rivalries and world wars."

Where did America fit in all this? Neither Ezekiel nor John nor any of the other prophets answered specifically, but, given NATO and other manifestations of trans-Atlantic cooperation, it appeared that the United States would be part of the kingdom of the west. Americans would certainly be involved in the stupendous clash that would take place when the four kingdoms met in combat, centered on Jerusalem. The carnage would be tremendous. "The apostle John predicts that so many people will be slaughtered in the conflict that blood will stand to the horses' bridles for a total distance of 200 miles northward and southward of Jerusalem." (Lindsey seems to have taken this particular prophecy figuratively, since he allowed that modern soldiers would use modern means of transport. Blood to the tops of the tank treads, perhaps.) Americans, at least those who got right with God, would also participate in the final victory. "As the battle of Armageddon reaches its awful climax and it appears that all life will be destroyed on earth—in this very moment Jesus Christ will return and save man from self-extinction."

On the surface, Lindsey's book had little to say about international relations in the present. His advice to readers was religious rather than political: Accept Jesus as your personal savior at once, so that when Jesus comes again he will class you with the sheep instead of the goats. Lindsey spent more space denouncing drug abuse, witchcraft, and the World Council of Churches than decrying detente, disarmament, or the Trilateral Commission.

Yet Lindsey's overall message reinforced the broad value system that informed the revival of the Cold War. His explicit identification of Russia as Gog might have been the text from which Reagan took his evil-empire views. His castigation of internationalism—referring to the story of the tower of Babel, he claimed, "This passage shows that God's plan for the world until the Prince of Peace returns is not an international one-world government, but nationalism"—fueled the arguments of those who attacked the United Nations, and who blasted efforts to substitute international law for armed force as the arbiter of disputes. His prediction of an unavoidable World War III abetted the conservatives' campaign to build up the American military. Lindsey didn't quite say that God was a Cold Warrior, but he indicated that the Deity as sure as heaven wasn't a detentist.[49]

CHAPTER SEVEN

Who Won the Cold War?
1984–1991

The limits of belligerence

Americans like strong characters, but by the end of the movie they usually prefer to see the hero smile and soften a little. They want their leads to be tough when conditions require, when the bad guys allow no other choice, but they like to know that good ultimately triumphs, and that the hero can eventually let down his guard.

By 1984, Reagan had been playing the tough guy for three years. With an election approaching, he and his handlers decided to lighten up. The president commenced the campaign season with a reassessment of where America stood with respect to the Soviets. "We've come a long way," he declared, "since the decade of the 1970s—years when the United States seemed filled with self-doubt and neglected its defenses, while the Soviet Union increased its military might and sought to expand its influence by armed force and threats." The ground covered was the result, naturally, of the efforts of the present administration. "Three years ago we embraced a mandate from the American people to change course, and we have. With the support of the American people and the Congress, we halted America's decline. Our economy is now in the midst of the best recovery since the 1960s. Our defenses are being rebuilt. Our alliances are solid, and our commitment to defend our values has never been more clear." Though the Soviets, who had expected the Americans to wallow in their weakness, initially had not known what to make of the American turnaround, by now the Kremlin was getting the message. This was good for the United States and for the peace of the world. "One fact stands out: America's deterrent is more credible, and it is making the world a safer place—safer because now there is less danger that the Soviet leadership will underestimate our strength or question our resolve."

Having regained its feet, the United States must move forward. "Deterrence is essential to preserve peace and protect our way of life," Reagan said, "but deterrence is not the beginning and end of our policy toward the Soviet Union." American leaders must establish a dialogue with the Soviet government, with the goal of achieving a constructive working relationship between the two countries. Whatever ideological differences divided the Soviet Union from the United States, the peoples of the two countries shared an interest in avoiding war and reducing the level of armaments.

Reagan cited three areas of relations as being both needful and susceptible of improvement. The first involved regional conflicts in such places as the Middle East, South and Southeast Asia, Central America, and Africa. The president conceded that most of these conflicts had originated in local disputes, and he doubted that the United States and the Soviet Union could terminate them. But he believed that Washington and Moscow could undertake "concrete actions" to reduce the risk that local conflicts would spread and suck in the superpowers.

The second area of necessary and possible improvement was the global arms race. "It is tragic to see the world's developing nations spending more than $150 billion a year on armed forces—some 20 percent of their national budgets." (Why it wasn't tragic to see the United States by itself spending more than $225 billion on armed forces—some 34 percent of the American national budget—the president didn't explain.) The United States and its allies had agreed to remove thousands of nuclear weapons from Europe. (The president declined to point out that many of these weapons were being replaced by more powerful weapons, such as Pershing II missiles.) "But this is not enough. We must accelerate our efforts to reach agreements that will greatly reduce nuclear arsenals, provide greater stability, and build confidence."

The third area was the general tone of Soviet-American communications. Exchanging insults would get the world nowhere. Washington and Moscow should seek to establish "a better working relationship with each other, one marked by greater cooperation and understanding." Reagan granted, albeit obliquely, that he had sometimes spoken of the Soviet Union in language that was less than diplomatic. "I have openly expressed my view of the Soviet system. I don't know why this should come as a surprise to Soviet leaders, who've never shied from expressing their view of our system." But an inclination to frankness needn't prevent the two sides from dealing fruitfully with each other. "We don't refuse to talk when the Soviets call us 'imperialist aggressors' and worse, or because they cling to the fantasy of a communist triumph over democracy. The fact that neither of us likes the other's system is no reason to refuse to talk. Living in this nuclear age makes it imperative that we do talk." The Soviets recently had broken off discussions on limiting intermediate-range nuclear forces in Europe, and had refused to set a date for a previously expected new round of talks on strategic and conventional weapons. (The president neglected to mention that the Soviet walkout had followed the deployment of the first Pershing IIs, which, as the Kremlin complained, could reach Moscow, while Soviet intermediate missiles couldn't reach Washington.) The Soviets' refusal to talk was regrettable, Reagan said, and should be corrected. "Our negotiators are ready to return to the negotiating table to work toward agreements," he added. "We will negotiate in good faith. Whenever the Soviet Union is ready to do likewise, we'll meet them halfway."

Actually, the administration had no intention of meeting the Soviets halfway, and, as events proved, all the significant concessions of the Reagan years came from the Kremlin. Nor did the speech offer anything Soviet leaders could hang their hats on in the other areas Reagan specified. Yet if the speech lacked substance, its style marked something of a turning point in Reagan administration policy. The president apparently had decided he had played the get-tough-with-Moscow role for about everything it was worth. Now he was adopting a more conciliatory approach. "Our two countries have never fought each other," he said (eliding the American intervention in Russia at the end of World War I). "There is no reason why we ever should." The Reagan of 1981 had thought he knew a very good reason why the United States and the Soviet Union might fight: because the Soviets were bent on conquering the world. The Reagan of 1984—"a year of opportunities for peace," he said—left the world-conquest paragraphs on the editing-room floor. He spoke to Soviet leaders, not as enemies of humanity, but as potential partners in the quest for human betterment. "Together we can strengthen peace, reduce the level of arms, and know in doing so we have helped fulfill the hopes and dreams of those we represent and, indeed, of people everywhere. Let us begin now."[1]

Even if 1984 hadn't been an election year, Reagan would have felt pressure to moderate his confrontational posture of the previous three years. The governments of the European allies had largely resisted the return to the Cold War, and NATO meetings during the early 1980s repeatedly turned into wrangling sessions, with the American delegates attempting to write a desire for detente out of joint communiques, and the Europeans attempting to write it in. The Europeans generally won, though not without allowing the Americans to insert the qualifier "genuine" before the offending noun. Other manifestations of inter-allied annoyance, such as the quarrel over the Soviet gas pipeline, similarly indicated that detente would go to its grave in Europe, if in fact it did, more slowly and far less quietly than it had in America.

While the governments of the allies experienced difficulty accepting Washington's reversion to Cold War form, opposition parties on the far side of the Atlantic had even more trouble. The British Labour party and West Germany's Social Democrats found it convenient to take Reagan at his bombastic word, and the image of the gun-slinging cowboy became a staple at rallies of the West European socialists. The Labourites flirted with unilateral nuclear disarmament—an idle threat at the time, considering Labour's political unpopularity (traceable in part precisely to this threat), but a position that indicated the widespread dissatisfaction with the worsening state of superpower relations. Like Labour, the German Social Democrats opposed the deployment of the American Pershing IIs and Tomahawk cruise missiles, with both parties complaining that the new weapons would place Soviet nuclear forces on even hairier triggers than at present, and would, in the bargain, make the launching sites of the missiles prime targets for Soviet rockets. Though Margaret Thatcher's Conservatives and Helmut Kohl's Christian Democrats overruled the opposition and ordered deployment to commence, the demonstrations and other disruptions that surrounded the deployment caused it to be a sensitive political topic for both London and Bonn.

The European anti-nukers had plenty of company in the United States. Carter's demise and Reagan's ascendancy had driven American detentists into the shadows, with elected officials and persons aspiring to office voicing broad support for a big defense budget and other signs of Cold War enthusiasm. But individuals and groups with no particular desire to stay within the tightening bounds of orthodoxy preached an alternative vision. Some opponents of Reagan administration policies were past the age of caring what was currently fashionable in Washington's power circles. In 1982, *Foreign Affairs* ran an article by former government officials George Kennan, McGeorge Bundy, Robert McNamara, and Gerard Smith advocating a shift in American nuclear policy to one of no-first-use. Ever since the early Cold War, the United States and its NATO allies had reserved the right to use nuclear weapons to counter a conventional offensive by the communists. This policy reflected the same influences that had given rise to Eisenhower's New Look, namely, a desire to defend Western Europe on the cheap. By adhering to a possible-first-use policy, the Reagan administration was breaking no fresh ground. Yet the bellicose tenor of various administration statements had alarmed Kennan and company, who worried that much of the world was getting the impression that a gang of warmongers controlled America's nuclear arsenal. By promising not to be the first country to hit the button, the United States would reassure those who wished to think better of America. At the same time, a no-first-use policy would reinforce global security by keeping the conceptual firebreak between conventional and nuclear weapons free of the clutter of improbable plans for limited nuclear war, which some persons associated with the Reagan administration had been talking about. "It is time to recognize that no one has ever succeeded in advancing any persuasive reason to believe that any use of nuclear weapons, even on the smallest scale, could reliably be expected to remain limited," the four authors stated. "Any use of nuclear weapons in Europe, by the Alliance or against it, carries with it a high and inescapable risk of escalation into general nuclear war which would bring ruin to all and victory to none."[2]

Less authoritative voices than those of Kennan, Bundy, McNamara, and Smith swelled the anti-nuclear chorus. Journalist Jonathan Schell provided *New Yorker* readers a graphic description of what a nuclear bomb could do to Manhattan. "Burst some eighty-five hundred feet above the Empire State Building," Schell wrote, "a one-megaton bomb would gut or flatten almost every building between Battery Park and 125th Street. . . . The physical collapse of the city would certainly kill millions of people. The streets of New York are narrow ravines running between the high walls of the city's buildings. In a nuclear attack the walls would fall and the ravines would fill up. The people in the buildings would fall to the street with the debris of the buildings, and the people in the street would be crushed by this avalanche of people and buildings." Schell went on to describe in detail the sequence of events following the detonation.

> A dazzling white light from the fireball would illuminate the scene, continuing for perhaps thirty seconds. Simultaneously, searing heat would ignite everything flammable and start to melt windows, cars, buses, lampposts, and everything else made of metal or glass. People in the street would immediately catch fire, and would shortly be reduced to heavily charred corpses. . . . Soon huge, thick clouds of dust and smoke would envelop the scene, and as the mushroom cloud rushed overhead (it would have a diameter of about twelve miles) the light from the sun would be blotted out, and day would turn into night. . . . Before long, the individual fires would coalesce into a mass fire, which, depending on the winds, would become either a conflagration or a firestorm. In a conflagration, prevailing winds spread a wall of fire as far as there is any combustible material to sustain it; in a firestorm, a vertical updraft caused by the fire itself sucks the surrounding air in toward a central point, and the fires therefore converge in a single fire of extreme heat. . . . In this vast theater of physical effects, all the scenes of agony and death that took place at Hiroshima would again take place, but now involving millions of people rather than hundreds of thousands.

There was much more along these lines, but this passage conveyed Schell's point. He concluded that the horrendousness of nuclear war, together with humanity's historical inability to resist using whatever tools of destruction it had devised, made the elimination of nuclear weapons morally imperative. "If we are honest with ourselves we have to admit that unless we rid ourselves of our nuclear arsenals a holocaust not only *might* occur but *will* occur—if not today, then tomorrow; if not this year, then the next.[3]

What Schell and the *New Yorker* did for the literate, ABC television did for the visuate. The network's *The Day After* depicted the consequences of a major nuclear attack on the United States. The special effects were amateurish compared with what Hollywood had accustomed American moviegoers to, but the storyline—America turned to rubble, American society shattered, the American people groping through the radiating ruins to regain a semblance of what they had taken for granted the day before—was chillingly potent. Although the program's producers eschewed any such explicit political statement as Schell's call for global disarmament, the show nonetheless provided powerful backing for arms-controllers, nuclear-freezers, and others opposed to the continuing buildup of the superpower arsenals.

Predictably, the Reagan administration's supporters rejected the anti-nuclear argument. The neoconservatives poured scorn on what they judged the simplemindedness of the no-nukers. Charles Krauthammer scored Schell for forgetting the accomplishments of nuclear deterrence. "Deterrence has a track record. For the entire postwar period it has maintained the peace between the superpowers, preventing not only nuclear war but conventional war as well." Krauthammer censured the nuclear-freezers for advocating what amounted to unilateral disarmament. Their pressure could prevent the United States and the Western allies from updating their arsenals, but who would enforce the freeze on the Soviets and the Chinese? The freeze movement played on people's fears of nuclear war, but it offered nothing constructive to substitute for the current system of great-power relations. "The freeze is not a plan; it is a sentiment," Krauthammer said. "The freeze continually fails on its own terms. It seeks safety, but would jeopardize deterrence; it seeks quick action, but would delay arms control; it seeks real reductions, but removes any leverage we might have to bring them about."[4]

Krauthammer likewise derided the no-first-use counsel of Kennan et al. Dubbing the four authors the "auxiliary brigade of the antinuclear movement," Krauthammer contended that unless a no-first-use policy were accompanied by a politically implausible upgrading of conventional forces, it would mean "the end of the Western alliance and the abandonment in particular of West Germany to Soviet intimidation and blackmail." Though the "four wise men"—Krauthammer made plain that he used the phrase ironically—acknowledged the need for conventional strengthening, they didn't give this need the emphasis it deserved. The hoi polloi of the anti-nuclear movement could be counted on to ignore it entirely. Irresponsible demagogues would appropriate the prestige of the four, and attach it to the disarmament agenda. Kennan and the others should have seen that this would happen. Hence, their raising of the first-use issue was perverse and dangerous. "The result of their highly publicized, grossly imbalanced proposal is predictable: another support in the complex and highly vulnerable structure of deterrence has been weakened. The world will be no safer for it."[5]

New kid on the bloc

Despite the neoconservatives' spirited defense of deterrence, of peace-through-strength, and of the Cold War, the Reagan administration continued to gravitate toward accommodation with Moscow. By the middle of 1984, Reagan's re-election was in the bag, and the president, like most White House second-termers, especially those of advanced years, began looking to the history books. Eisenhower, in the late 1950s, had taken the easing of superpower tensions to be his primary goal—a quest facilitated by the departure of John Foster Dulles, who died in 1959. In Reagan's case, the shift to a less confrontational posture was similarly eased by a change at the State Department. George Shultz wasn't less disposed than Alexander Haig, Reagan's first secretary of state, had been to use military force when conditions required, but Shultz was considerably less disposed than General Haig to make combative statements and martial gestures. In October 1984, Shultz outlined his understanding of the appropriate method for dealing with the Soviets. Shultz gave the necessary nod to the "profound differences" between the American and Soviet approaches to world affairs, saying that the American government, embodying the will of the American people, respected the rights of other peoples to pursue their legitimate objectives undisturbed, while the Soviet government, embodying a totalitarian ideology, attempted to impose its will on other countries and peoples. But then Shultz got to the meat of his message. "Despite these profound differences, it is obviously in our interest to maintain as constructive a relationship as possible with the Soviet Union." For better or worse, the Soviet Union wouldn't soon disappear. Nor did Shultz think this was entirely for the worse. "Its people are a great and talented people, and we can benefit from interchange with them." In any event, the United States couldn't ignore them or their government.

The essential issue, Shultz declared, was the degree to which Soviet bad behavior in one area of relations required American sanctions elsewhere. Shultz didn't entirely reject linkage of this sort, but he thought it easy to overdo. "Linkage as an instrument of policy has limitations," he asserted. "If applied rigidly, it could yield the initiative to the Soviets, letting them set the pace and the character of the relationship." The secretary of state reminded listeners that American negotiators didn't negotiate for the fun of it. "We negotiate when it is in our interest to do so." For this reason, the United States would err to break off negotiations on one topic to register disapproval of Soviet policy on another. The Carter administration had fallen into this error after the invasion of Afghanistan, and neither Americans nor Afghans had benefited. (This was quite a statement coming from a top Reagan official—criticizing Carter for being too tough on the Russians.) Shultz quoted Winston Churchill on the subject of linkage: "It would, I think, be a mistake to assume that nothing can be settled with the Soviet Union unless or until everything is settled." (Shultz didn't add that Eisenhower and Dulles had rejected Churchill's argument and vetoed the British prime minister's proposal for a summit with the new post-Stalin leadership.)

Some persons contended that negotiations with the Kremlin would forever fail because the Soviets could never be trusted. "But the truth is," Shultz countered, "successful negotiations are not based on trust. We do not need to trust the Soviets; we need to make agreements that are trustworthy because both sides have incentives to keep them." Shultz also rejected the argument that the United States should keep building weapons until negotiations became unnecessary. "Our premise is that we should become strong so that we are *able* to negotiate." Progress would take time, but it was definitely possible. "The way is wide open to more sustained progress in U.S.-Soviet relations than we have known in the past."[6]

One reason Shultz adduced for his optimism was a recent remark by Soviet party boss Konstantin Chernenko (who succeeded Yuri Andropov, who had followed Brezhnev briefly), that Moscow likewise desired a constructive dialogue. Perhaps Chernenko's actions would have justified Shultz's optimism. But the Soviet leader didn't live long enough for the world to find out. In March 1985, politburo primacy passed to Mikhail Gorbachev, a younger and much more energetic man. Under the twin rubrics of *glasnost* (openness) and *perestroika* (restructuring), Gorbachev swiftly set out to revolutionize Russian domestic and foreign policy as no one had since Lenin.

Gorbachev's impact on American foreign policy was hardly less. By instigating the remaking of the Soviet Union, and by allowing the remaking of Eastern Europe, Gorbachev challenged the assumptions on which American Cold War policy had rested for forty years. In a stunningly short period, he deprived Americans of the only major enemy most of them had ever known, and, like the half of a two-person tug-of-war who unwarningly lets go, he threw America's Cold War apparatus abruptly off balance. He forced American leaders and the American people to devise new definitions of American national interests, and to design new methods of securing these interests. It was arduous work, and would take time.

The Reagan administration responded to Gorbachev's reforms by pursuing its policy of accommodation toward the Soviet Union. For decades, American representatives had complained of Moscow's slowness in negotiations. The Russians carried patience to the point of psychological warfare, wearing down their interlocutors not by argument but by tedium. Gorbachev reversed the situation overnight, firing off a salvo of new proposals, which he followed up with a parade of ideas during subsequent months. For starters, he offered to extend the 1963 partial nuclear test ban to a prohibition of all nuclear testing, underground as well as in the atmosphere. He announced a unilateral Soviet moratorium on tests, suggesting that Washington might respond in kind. He forwarded a plan to reduce the strategic nuclear arsenals of the superpowers by 50 percent, in exchange for an agreement by the United States to stop work on strategic defenses and for other lesser concessions.

The unaccustomed motion in Moscow caught the Reagan administration by surprise. The Defense Department did what the Kremlin had done for years when confronted with the unexpected: it said no, and later provided reasons. The opposition by Pentagon officals to Gorbachev's new approach reflected a variety of considerations. In the first place, the Pentagonists suspected a trick, on the zero-sum thinking that anything the

Russians would suggest must, ipso facto, be bad for the United States. In the second place, even if the offers contained no hidden booby traps, the Soviets seemed to be in a compromising mood. The United States should hold firm and see what Moscow would offer next. Third, the specific offers put on the table thus far would work to America's peculiar disadvantage. The United States relied on nuclear weapons to offset the Soviet edge in conventional forces. A test ban would inhibit America's ability to maintain a high condition of reliability in present weapons, and to develop more-effective future weapons. As for strategic defense, the United States evidently held an advantage in this crucial area of military technology. It would be foolish to relinquish the advantage.

What went without saying, and perhaps even without thinking among the truest believers in the Pentagon, was that the kinds of curbs the Kremlin suggested would cut seriously into the military buildup that was making the generals, admirals, and defense contractors happier than they had been since the flush times of the early 1950s. Whether or not arms control would stabilize world affairs, it would *de*stabilize, bureaucratically and economically, all those involved in the operation of the world's largest purchasing organization. A test ban would slow or halt the development of the next generation of nuclear weapons, thereby slowing or halting the development of the next generation of military officers, civilian government officials, scientific researchers, business executives, union leaders, and production workers employed in the various phases of the weapons' progress from design to deployment. A reduction in nuclear strategic forces would lead to a reduction in the administrative and labor battalions the nuclear forces required. A shelving of SDI would slam the door on the most promising project the military-industrial-scientific complex had seen since the atom bomb.

The Pentagon took extraordinary measures to prevent the president from accepting Gorbachev's new thinking. On the eve of a November 1985 Geneva summit meeting, Caspar Weinberger sent Reagan a letter warning against the dangers the summit entailed. At this stage, the State and Defense departments were arguing about how far America should continue to honor previous arms-control agreements and near-agreements, in the face of alleged Soviet violations. The defense secretary cautioned the president:

> In Geneva, you will almost certainly come under great pressure to do three things that would limit severely your options for responding to Soviet violations. The first is to continue to observe SALT II. The second is to agree formally to limit SDI research, testing and development to only the research allowed under the most restrictive interpretation of the ABM treaty, even though you have determined that a less restrictive interpretation is justified legally. The Soviets doubtless will seek assurances that you will continue to be bound to such tight limits on SDI development and testing that would discourage the Congress from making any but token appropriations. Third,

the Soviets may propose communique or other language that obscures their record of arms control violations by referring to the "importance that both sides attach to compliance."

The president must dodge these dangers, Weinberger declared. "Any or all of these Soviet proposals, if agreed to, would sharply restrict the range of responses to past and current Soviet violations available to us."[7]

Perhaps to ensure that the president paid proper attention to this advice, someone with access to the letter leaked it to the press. Not surprisingly, the maneuver did not conduce to profitable negotiations with Gorbachev, which was precisely the point. White House aides were furious. One top-level official, asked by a reporter if the Weinberger letter had been "sabotage," replied, "Sure it was."[8]

Much of the reason for the intra-administration sniping—aside from the fact that the State Department and the National Security Council lacked the Defense Department's proprietary interest in a big army, navy, and air force—was that the president hadn't made up his mind about arms control. Some question existed whether Reagan had the essential issues entirely in hand. Comments spoken in all apparent sincerity suggested either that the president was confused about the relationship between SDI and other weapons systems, or that he was playing an exceedingly deep game. At the end of October, he told a group of Soviet journalists that the deployment of SDI would come only after the elimination of offensive missiles. As he phrased it, "I have said and am prepared to say at the summit that if such a weapon is possible, and our research reveals that, then our move would be to say to all the world, 'Here, it is available.' We won't put this weapon, or this system, in place, this defensive system, until we do away with our nuclear missiles, our offensive missiles." These remarks sent shock waves across the Potomac to the Pentagon, where they raised the specter of an easy Soviet veto of SDI, accomplished merely by refusing to eliminate offensive weapons.[9]

A week later, after some frantic re-briefing by defense officials, the president clarified his position—or muddied it, depending on one's point of view. He denied what certain persons had inferred from his earlier statement. Speaking of SDI, he said, "Someone just jumped to a false conclusion when they suggested that I was giving a veto to the Soviets over this." The United States wouldn't let Soviet stonewalling stop deployment.

> Obviously, if this took place, we had the weapon—I keep using that term; it's a defensive system—and we could not get agreement on their part to eliminate the nuclear weapons, we would have done our best, and, no, we would go ahead with deployment. But even though, as I say, that would then open us up to the charge of achieving a capacity for a first strike. We don't want that. We want to eliminate things of that kind. And that's why, frankly, I think that any nation offered this under those circumstances that I've described would see the value of going forward. Remember that the Soviet Union has already stated its wish that nuclear weapons could be done away with.[10]

Another week later, Reagan further refined his position. Asked whether he still intended to share the results of SDI research with the Russians, the president said,

> Maybe I didn't make it clear. That's what I meant in my earlier answer—not just share the scientific research with them. Let me give you my dream of what would happen. We have the weapon. We don't start deploying it. We get everybody together, and we say, "Here, here it is. And here's how it works and here's what it'll do to incoming missiles." Now, we think that all of us who have nuclear weapons should agree that we're going to eliminate the nuclear weapons. But we will make available to everyone this weapon. I don't mean we'll give it to them. They're going to have to pay for it—[laughter]—but at cost. But we would make this defensive weapon available.[11]

Whatever uncertainty his pre-summit remarks may have sown, by the time Reagan got to Geneva, he had his position straight. He wouldn't abandon work on strategic defense. "I simply cannot condone the notion of keeping the peace by threatening to blow each other away," he told Gorbachev. "We must be able to find a better way." Reagan explained that the United States didn't intend for SDI to destabilize deterrence, nor to provide a cover for achieving nuclear superiority. He reiterated his desire to share defensive technology with other countries, and he promised to allow Soviet scientists access to American research facilities, in order to verify the defensive nature of the work going on there.

Gorbachev wasn't buying. The general secretary contended that though the Americans might call SDI purely defensive, it could just as easily yield "offensive nuclear weapons circling the earth." The talk of sharing technology was so much eyewash. The United States was plotting, as it had plotted repeatedly in the past, to achieve a "one-sided advantage" over the Soviet Union. Responding to Reagan's reminder that the United States hadn't abused its nuclear monopoly after World War II, and to the president's question, "Why don't you trust me?," Gorbachev asked why Reagan didn't trust *him*. When Reagan said that as American president he had an obligation to take into account the capabilities of the Soviet Union, as well as the intentions of its leaders, Gorbachev countered that as Soviet general secretary he himself could be no less vigilant. Whatever America's present intentions, Gorbachev said, SDI would give the United States enormous destabilizing capabilities. The Soviet government could never allow the threat of such destabilization to arise.[12]

In the end, the Geneva summit produced nothing beyond handshakes and smiles. Reagan held to SDI tighter than ever, and the anti-arms-controllers in the Pentagon breathed more easily.

Yet Gorbachev hadn't run out of ideas. In January 1986, the Soviet leader laid out a breathtaking scheme for nothing less than the total elimination of nuclear weapons by the end of the century. The scheme amounted to an attempt to call the Americans' bluff. Reagan had been plumping SDI as a means for making offensive nuclear weapons obsolete, and now Gorbachev was going the president one better by offering to make them not merely obsolete but nonexistent. Needless to say, the proposal was principally a propaganda ploy, since no one seriously expected that either superpower would agree to reduce itself militarily to the status of an India or a Brazil. All the same, the proposal threatened to steal much of Reagan's peacekeeping thunder.

Washington had dealt with Soviet propaganda before, and it was not totally unprepared for this round. Yet a subsidiary part of Gorbachev's proposal occasioned considerable concern. At the time when the Reagan administration had been attempting to persuade the Europeans to accept the Pershing IIs and Tomahawks, American officials—led by Assistant Defense Secretary Richard Perle—had offered what came to be known as the zero option. Under this scheme, the United States and its allies would forgo deployment of the Pershings and Tomahawks, in exchange for the Soviet Union's dismantling of its own SS-20 intermediate-range missiles. The result would be no NATO intermediate missiles and no Soviet intermediate missiles. Few persons on either side took the zero option at face value. The Soviets rejected it because it required them to trade their own missiles already in place for missiles the Americans hadn't deployed and might never. The Americans knew that the Soviets would reject it, and some, like Perle, were happy they did. Perle's objective was not so much to get the SS-20s out of Europe as to get the Pershings in. After all—as the Soviets endlessly complained—the Pershings targeted Moscow, while the SS-20s targeted merely Bonn, Paris, and London. The basic aim of the zero option was to provide political plausibility to the position of the European governments who wanted to deploy the American missiles, but had to deal with vocal opposition groups who didn't.

By 1986, when hundreds of the American intermediate missiles had been placed in Europe, the zero option had gained substantial appeal for Moscow, and lost a great deal for Washington. Now a zero solution would involve trading real threats for real threats, rather than real for hypothetical. As a consequence, when Gorbachev announced that the Soviet Union accepted the zero principle, the Reagan administration began backing and filling. Though the zero option originally had focused on Europe, American officials reminded everyone listening that the plan had called for the elimination of Soviet SS-20s in Asia as well, since these mobile missiles could be moved to Europe in a crisis. The Pentagon tried to divert attention from the now relatively realistic zero option for intermediate nuclear forces, by pushing a proposal for outlawing all ballistic missiles—a measure as unrealistic and therefore safe as Gorbachev's scheme for the total elimination of nuclear weapons.

Simultaneously, the Soviets continued to pound away at SDI. They denounced it as a new form of warmongering, and productive only of an unprecedentedly expensive round of the arms race. A growing number of

Americans came to agree. Congress, realizing that the budget deficit wouldn't vanish on its own, and recognizing that the president was now a lame duck, began to question whether the country could afford a defensive system judged impractical by some of America's best scientists. In addition, the fresh image of Soviet leadership Gorbachev was cultivating—reasonable, non-threatening, more interested in domestic reform than foreign adventures—was having its desired effect. Brezhnev had made a convincing bogey, but Gorbachev didn't.

In response to the growing opposition, SDI's American supporters launched a counterstrike. A group denominating itself the Coalition for SDI warned the president against bargaining away America's advantage in strategic defense. Another organization, the Center for Peace and Freedom, decried the danger of an "SDI sellout." The so-called "laser lobby" on Capitol Hill pressed for a commitment not simply to SDI research but to full development and deployment. Congressman Jack Kemp, whose previous support for the zero option was returning to haunt him, especially now that he was preparing a run for the 1988 Republican presidential nomination, complained publicly about persons in the State Department who were trying to get Reagan to step back from SDI. Matters reached such a pass that Paul Nitze found himself branded a dove for expressing less than unrestrained enthusiasm about SDI. Nitze could hardly believe what he heard and read. "It gets me down," he told a friend, "to be identified as a giveaway artist."[13]

Reagan tried to calm the fears that he would let the Soviets come between the star warriors and their baby. Replying to reports that in a letter to Gorbachev he had suggested a "grand compromise" offering to swap SDI for major reductions in offensive weapons, the president told a news conference, "Let me reassure you right here and now that our response to demands that we cut off or delay research and testing and close shop is: No way. SDI is no bargaining chip; it is the path to a safer and more secure future. And the research is not, and never has been, negotiable." Reagan slammed opponents of SDI for playing into Soviet hands. Efforts in Congress to curtail SDI funding "could take away the very leverage we need to deal with the Soviets successfully." Persons who claimed that SDI wouldn't work suffered from "clouded vision." "Sometimes politics gets in your eyes," the president said. If SDI was as big a waste as opponents claimed, why were the Soviets so interested in blocking it? Closing his argument with an anecdote, he told how Robert Fulton had tried to sell his steamboat to Napoleon. The emperor had scoffed at the notion of boats defying wind and current, Reagan said. "Let's not make the same mistake."[14]

Gorbachev guessed that he might have better luck dealing with the president personally than working through the divided American bureaucracy. He suggested a tête-à-tête. Reagan and Gorbachev already had a summit slated for Washington in December, but the general secretary thought a quiet weekend without all the summit fanfare might facilitate cooperation. How about Iceland in October? Gorbachev doubtless appreciated that a pre-election summit would appeal to Reagan. The Geneva meeting of 1985 had given the president's ratings a boost, and with the Republicans suffering from Americans' chronic sixth-year political itch, Gorbachev believed that Reagan would be glad for anything that might help his party hold the Senate. Reagan was.

American officials hadn't yet gotten used to Gorbachev's style of negotiation, and his actions at Reykjavik again took the administration by surprise. Reagan and his advisers flew to Iceland chiefly prepared to talk about the intermediate nuclear forces, but at the first session, the general secretary suggested that the two leaders think big, that they "wrench arms control out of the hands of the bureaucrats." Gorbachev proceeded to describe what the wrenching ought to entail. He once more called for across-the-board cuts of 50 percent in strategic offensive weapons, but this time specified that the cuts should include "substantial—I don't mean trivial, but substantial" reductions in heavy missiles, the ones the United States deemed so threatening. He offered to accept American definitions of strategic weapons (a previously sticky point), and he dropped the condition that British and French strategic weapons be counted as part of the American stockpile. He restated his acceptance of the zero option, indicating that a solution to the European-versus-Asian aspect of the issue could be worked out. He said he'd allow the American SDI project to proceed, subject only to a pledge that neither side would overturn the ABM treaty for ten years.

Over that night, Gorbachev's and Reagan's spear-carriers tussled about the precise form the proposed cuts should take, and about exactly what the ABM treaty said regarding anti-missile defenses. They came close to agreement, and their work laid the basis for the eventual accord on intermediate nuclear forces. But the talks stuck on SDI, specifically on whether SDI research had to take place within laboratories. The Soviets wanted to keep SDI indoors, while the Americans wanted the freedom to try out promising ideas in the atmosphere or in space.

At the final session of Reykjavik meeting, Gorbachev attempted to break the deadlock by reiterating his desire for total de-nuking. "I would favor eliminating all nuclear weapons," the general secretary declared. Reagan perked up. That was what he had been aiming for all along, the president replied. "Then why don't we agree on it?," Gorbachev asked. "Suits me fine," Reagan said.

Then came the hook. "But this must be done in conjunction with a ten-year extension of the ABM treaty and a ban on the development and testing of SDI outside the laboratory," Gorbachev said.

In subsequent weeks, Reagan's agreement to the elimination of nuclear weapons occasioned a minor controversy. Was the president speaking in general terms of a nuclear-free world, or in particular of the kind of ten-year timetable Gorbachev proposed? Administration officials assured worried European leaders, who cringed at the thought of facing all those Soviet tanks without an American nuclear shield, that the president was merely expressing a someday wish for humanity.

One thing Reagan's assent clearly did *not* encompass was the required constraint on SDI. "There is no way we are going to give up research to find a defense weapon against nuclear missiles," Reagan told Gorbachev. On this rock the meeting broke up. "We tried," Paul Nitze told journalists as the president and the general secretary headed for their planes. "By God, we tried. And we almost did it." But only almost.[15]

Because it's there—and Ollie's not

Reagan returned from the Reykjavik meeting and walked smack into the Iran-contra scandal. Not much connected the two events causally, although many of the same people were involved. But each contributed to the de-escalation of the Cold War. While the Iceland summit failed to produce the kind of sweeping arms control agreement the two principals batted around during their weekend at Hofdi House, it demonstrated definitively that, in Gorbachev, the Soviets had a leader willing and able to deal. It also suggested that the hard-liners in the Reagan administration wouldn't be able to block arms control forever. Gorbachev was saying yes faster than they could think of reasons to say no, and at some point they would have to accept victory.

The Iran-contra affair eroded the foundation of the Cold War from another direction. Previous presidents had crossed the line into activities that most Americans, in most circumstances, probably would have considered unethical or immoral, such as trying to assassinate Castro and Lumumba. But because Congress had granted the executive branch great discretion in waging the Cold War, the dirty tricks involved (at least the ones played outside the United States) weren't illegal. The Iran-contra affair was different. The questionable pedigree and checkered performance of the Nicaraguan guerrillas had put off sufficient numbers of senators and representatives that Congress had decided to pull the plug on lethal American aid. But just as the Reagan White House wouldn't (yet) take yes for an answer from Moscow on arms control, it wouldn't take no from Congress on Nicaragua. It set about funding the contra war by extra-congressional means. Legal counsel to the administration's intelligence oversight board advised the president that though the Boland amendment prohibited the CIA and the Defense Department from funding the contras, the measure didn't apply to the National Security Council. Unsurprisingly, the administration declined to inform Congress of this advice. As the counsel, Bretton Sciaroni, circumspectly conceded to the Iran-contra committee later, "It would seem to be the implication that if Congress found out about the legal opinion, it would move to prevent NSC officials from acting." When word eventually surfaced that the Reagan administration might have violated the law, Congress had no choice but to investigate. The Democrats on Capitol Hill were happy to.[16]

The Iranian connection in the contra-supply business rendered an investigation even more necessary. If the administration hadn't been caught in the exceedingly compromising position of trying to barter arms to Iran for American hostages, and if the president hadn't been caught lying about the matter, the public might well have shrugged at the Nicaraguan side of the fiasco. Polls consistently showed a lack of strong feeling one way or the other about Nicaragua, and although the administration may have failed to heed congressional strictures on aid to the contras, the American people didn't exactly hold Congress in the highest esteem.

As matters turned out, the unearthings the Iran-contra committee accomplished did surprisingly little political damage to Reagan personally. Oliver North may or may not have masterminded the operation, but his televised flag-waving deterred the committee from delving too deeply into the affair. More to the point, most Americans didn't desire to see Reagan disgraced. He was a nice old man, entirely unlike that shifty Richard Nixon, whose Watergate peccadillos afforded the obvious parallel. Besides, after six years of a conspicuously casual style of leadership in the Oval Office, many Americans were willing to believe that the president really didn't know what some of his closest advisers were doing.

All the same, as the muck of the Iran-contra fiasco rose around his administration, Reagan couldn't help recognizing that further summiteering would tend to lift him above the mess. By the beginning of 1987, Gorbachev was the most attractive politician in the world—if not necessarily in the Soviet Union—and foreign leaders were lining up to be photographed with him. Reagan could reasonably hope that some of the good ink Gorbachev was receiving would rub off. Moreover, concluding a landmark arms-control agreement with the Soviet Union would go far toward making people forget the venial violations of the Iran-contra affair.

The path to an agreement was made easier by the departure of key Cold Warriors in the administration. Richard Perle, known around Washington as the "Prince of Darkness," both for his brooding visage and for his unrelenting opposition to arms control, announced his resignation in March. As he told a friend in leaving, it was "getting to be springtime for arms control around here," and he preferred to quit uncompromised. Kenneth Adelman, the director of the Arms Control and Disarmament Agency, and Perle's partner in opposition, bailed out a few months later. Caspar Weinberger, sensing a shift toward tighter budgets, decided that telling the generals and admirals they couldn't have everything they wanted would be less fun than telling them they could, and gave notice in November.[17]

An even more important roadblock to an INF agreement disappeared in February, when Gorbachev detached the issue of intermediate missiles from the question of star wars. The White House praised Gorbachev's action, and sent the pro-arms control George Shultz to Moscow to work out details. Some American officials still feared a trap. The soon-to-retire—but hardly retiring—American commander of NATO, General Bernard Rogers, warned of the dangers of a denuclearized Europe in which Warsaw Pact conventional forces would continue to outnumber those of the Atlantic alliance. Rogers went on to excoriate unnamed "pre-emptive conceders in high positions in the United States government" who were rushing to agreement chiefly for the sake of agreement. Shultz, on the verge of vanquishing the anti-arms-controllers, wasn't about to let a rogue general kill the INF deal. The secretary of state declared Rogers "way out of line," and asserted that his suggestion that an INF agreement would endanger American security was "entirely incorrect." Such an agreement would make America more secure, not less.[18]

From here, it was smooth sailing. In September, following additional conversations between Shultz and Soviet foreign minister Eduard Shevardnadze, Reagan announced tentative concurrence on an INF accord. The finalized treaty formed the centerpiece of a December Reagan-Gorbachev summit in Washington. A few bitter-enders in the Senate still found reason to oppose the treaty, but in May 1988 the upper chamber ratified the INF pact by a vote of 93 to 5, delivering its approval just in time for another summit, this in Moscow.

Reagan couldn't get enough of summitry. In all, the president met Gorbachev five separate times. This was more than any of his predecessors had met Soviet leaders, more than anyone would have guessed during Reagan's evil-empire phase, and far more than suited those conservatives and neoconservatives who still clung to the Cold War. By the last meeting, the two presidents (Gorbachev having been elevated to Soviet head of state) were best buddies. "Since our first summit in Geneva three years ago," Reagan said, "we've traveled a great journey that has seen remarkable progress, a journey we continue to travel together." The distance traveled since the early days of the Cold War was even greater. "The decades following World War II were filled with political tensions and threats to world freedom. But in recent years, we've seen hopes for a free and peaceful future restored and the chance for a new U.S.-Soviet relationship emerge." So taken was Reagan with what he and Gorbachev had accomplished that he described the current period as "the brightest of times."[19]

Is it over yet?

Reagan didn't quite consign the Cold War to the ash heap of history, but he came close. Gorbachev left him little choice. In February 1988, the Soviet leader announced that Soviet troops would begin to evacuate Afghanistan by May, and would complete their pullout by the following February. In March, he ordered his defense minister to get together with the American defense secretary to discuss military doctrine, with an eye to reducing superpower tension further. At the Moscow summit in May, he and Reagan concluded a variety of agreements for Soviet-American co-operation, covering activities ranging from nuclear-power research to the rescue of fishermen in distress. In September, he offered to close the Soviet naval base at Cam Ranh Bay in Vietnam, if the United States agreed to close its facilities in the Philippines. At the same time, he proposed to place the controversial Krasnoyarsk radar station (seen by some American experts as violating the ABM treaty) under international supervision, and to convert it to a space science station. In December, he ordered the dismantling of two other radar installations that had raised similar questions.

Meanwhile, throughout the year, the internal reforms Gorbachev had set in motion within the Soviet Union gained momentum. Dissenters spoke more freely, both inside the Communist party and outside it, and for the first time in their careers, party officials faced the prospect of competitive elections. *Glasnost* and *perestroika* began to spill over into Eastern Europe after Gorbachev encouraged the allies to design "innovative policies" to deal with the stagnation left over from the bad old days of his predecessors. Poland's government agreed to negotiations with Solidarity. Hungary sacked Janos Kadar and nearly half the country's Central Committee. Ten thousand Czechs gathered in Wenceslas Square to commemorate the "Prague spring" of 1968.

As the evidence accumulated that Gorbachev was a phenomenon fundamentally different from anything the Kremlin had housed since the October Revolution, some American commentators were more than happy to declare the Cold War over. George Kennan stated that containment was now "irrelevant." In fact, he said it had been irrelevant for quite some time. With the possible exception of the period of the Berlin blockade of 1948–49, the Soviet Union had never posed a serious military threat to the United States. The more important Soviet ideological challenge had largely disappeared after the European countries on the American side of the Elbe had regained their economic and political feet in the 1950s and 1960s. Recent developments within the Soviet Union indicated that the ideological challenge had vanished for good, since it was evident that not even the communists believed in their ideology any longer. Marxism-Leninism was "a stale and sterile ritual," Kennan wrote. Soviet leaders might pay lip service to the old icons for awhile yet, since the icons provided the sole legitimation for the leaders' continued hold on government. But commmunist ideology no longer moved the leaders or their country. The United States should bear this in mind in relations with the Soviet Union. "The communist aspect of it all has very little to do with the Soviet Union today."

Kennan had long criticized America's over-emphasis on military matters in dealing with Moscow, and now he denied that the military buildup of the 1980s had had much to do with the changes in the Soviet Union. If anything, he asserted, Washington's hard line had strengthened the hand of hard-liners in the Kremlin who opposed the changes Gorbachev was trying to implement. The mellowing of the Soviet system was "primarily the result of forces operating within Soviet society." The most important of these forces were disillusionment at the failure of communism to deliver the material benefits it had promised, disgust at the brutality of life under a form of government still operating by essentially Stalinist principles, and dissatisfaction among ethnic minorities at their subordination to the Russian majority. As the Soviet people grew increasingly aware of conditions outside their country, and of the gap that separated them from the advanced nations of the West, the more insightful Soviet leaders—outstandingly Gorbachev—were deciding that only major reform could prevent the Soviet Union from falling by the wayside.

If America's Cold War mindset had ever served a useful purpose, it no longer did. Kennan conceded that the American ideology of the Cold War had provided a psychological relief valve for that sizable segment of the American population that needed "to cultivate the theory of American innocence and virtue, which must have an opposite pole of evil." But it was time to grow up. "The extreme military anxieties and rivalries that have marked the high points of the Cold War have increasingly lost their rationale. Now they are predominantly matters of the past. The Cold War is outdated." The Soviet Union was ahead of the United States in recognizing this fact. Americans had better follow Russia's example. "The Soviets dropped the Cold War mentality. Now it's up to us to do the same thing."[20]

The neoconservatives disagreed. While granting that something was happening behind the Iron Curtain, the intelligentsia of the Reagan Cold War weren't ready to declare their services dispensable. Jeane Kirkpatrick remained to be convinced that events in the Soviet Union were proving her wrong about the distinction between totalitarianism and authoritarianism. "While there is evidence to suggest that totalitarian states are capable of change," Kirkpatrick remarked at the end of 1988, "there is so far no example of such a regime evolving into something different." Kirkpatrick admitted that communist ideology in the Soviet Union no longer possessed the persuasiveness it once had, but she held that the crumbling of communism, if such was what was at hand, didn't necessarily imply the end of totalitarian rule. "The Russian tradition of the theocratic state is a discouraging prologue to *perestroika*. Orthodox doctrine has been married to the sword for a very long time. Before communist ideology was joined with state power, the Czar and the Russian Orthodox Church were the omnipotent authority." Which way would Gorbachev jump next? Kirkpatrick, less confident predicting the future than she had been a decade earlier, said she awaited new developments with "rapt attention."[21]

Charles Krauthammer adopted a similar view. Krauthammer noted a statement by Britain's Margaret Thatcher—normally a neoconservative favorite—that the Cold War was over, and said, "Uncharacteristically, she is wrong. Such thinking is wishful until the Soviets leave not just Afghanistan but Central America, until they not only talk about 'defensive sufficiency' but practice it by making real cuts in defense spending and by reconfiguring their offensive force structure in Europe." Yet Krauthammer didn't claim that nothing had changed. Much had, and more might. "For the first time in the postwar period it is possible to foresee an end to the Cold War—on Western terms."

Krauthammer perceived two routes leading beyond the Cold War. One ran from totalitarianism to authoritarianism. (Krauthammer had less residual faith in the Kirkpatrick doctrine than the doctrine's author.) "It is conceivable that in the foreseeable future the U.S.S.R. will have been transformed into a merely authoritarian one-party state, not terribly more illiberal than most of the 19th century monarchies." In this case, its ideological engine out of fuel, the Soviet Union would conduct its foreign relations much like other major powers. "Not a guarantee of peace, by any means, but a respite from the prospect of unending, irreconcilable hostility"—in other words, though Krauthammer didn't utter them, detente as envisaged by Nixon and Kissinger.

The other possible route leading beyond the Cold War would traverse the wreckage of *glasnost* and *perestroika*. If Gorbachev's reforms failed, the Kremlin, probably under a successor regime, might desire to continue the Cold War, but would lack the capacity. A recognition of this fact, Krauthammer asserted, was precisely what had inspired the present changes. "The Soviets know that their historic achievement of the last forty years—being able to match a coalition of the greatest powers in history: the United States, Britain, France, Germany and Japan—is slipping out of their reach. Soviet rethinking on this score is not due to any great soul-searching. It is merely a response to objective reality." Regardless of the cause, the failure of reform would leave the Soviet Union worse off than ever. "The old system will likely not be able to sustain itself and certainly not be able to maintain a policy of imperial expansion."

Krauthammer held that with the Kremlin on the ropes, now wasn't the time to ease the pressure or declare the Cold War over. "Indeed, declaring it over or being willing to offer the other guy a draw is one way of blowing it." The Soviets were suing for peace, because they knew that otherwise they'd lose. "By challenging Soviet acquisitions, by leading a worldwide economic resurgence, by launching an arms race that the Soviets have had great trouble matching, the United States has convinced the Soviets that, if things continue, they can no longer win the twilight struggle." On the eve of victory, the United States must remain firm.

Krauthammer wasn't sure that victory, assuming it did come, would be an unalloyed blessing. "Nations need enemies," he wrote. "Take away one, and they find another." Why? For purposes of self-identification and motivation. "Parties and countries need mobilizing symbols of 'otherness' to energize the nation and to give it purpose." Examining the history of the Cold War, Krauthammer detected a disquieting tendency toward confusion in American foreign policy during periods when the national enemy hadn't been clearly defined. In the decade and a half after 1945, Americans almost unanimously agreed that the Soviet Union was America's chief and mortal enemy. The Vietnam War undermined this consensus, and American policy floundered. Americans after Vietnam lacked an obvious and appropriate outlet for the natural hostilities all persons feel. They turned their hostilities toward lesser evils, toward what Krauthammer called America's "ugly friends": Pinochet in Chile, Park in Korea, Marcos in the Philippines, Pahlavi in Iran, Somoza in Nicaragua. During this period of cognitive and moral disarray, the Soviets made significant advances, to the extent that the Kremlin could dream of winning the Cold War. But the Reagan-era "recovery from Vietnam" restored direction and perspective to American policy, and "abolished the brief communist fantasy of a Pax Sovietica."

Now, with the overly optimistic already declaring the Cold War ended, Americans had to face the problem of finding another enemy. Krauthammer worried that if the 1988 presidential campaign offered any guidance, Americans again would take to bashing their allies. Richard Gephardt had suggested economic sanctions against Japan, as a means of countering what he considered unfair trading practices. Other candidates had followed suit. Krauthammer judged this a very disturbing, though predictable, development.

Yet such problems remained the business of the future. At present, the United States must concentrate on winning—not just ending—the Cold War. "It would be a historic tragedy," Krauthammer concluded, "to settle for anything less than victory now that it is in sight for the first time."[22]

Zbigniew Brzezinski wasn't going to settle for less than victory either, although not from neoconservative skepticism (Brzezinski being no neoconservative), but from a belief that nothing less than victory was at all likely. The Cold War was as good as over, the former Carter administration hawk declared, because communism had entered its "terminal crisis." Within a short time, communism would beR"largely irrelevant to the human condition," and would hereafter be remembered primarily as "the twentieth century's most extraordinary political and intellectual aberration." Brzezinski described Gorbachev's reforms in terms of trying to remove three layers of communist dogma and practice. Removal of the first layer, that of the Brezhnev era, was well under way. Work on the second, Stalinist, layer had begun. The deepest, Leninist, layer so far remained beyond the reach of the chippers. Gorbachev hadn't conceded the need to tackle Lenin's legacy, and he would find it exceedingly difficult to do so, for Leninism provided the foundation of Communist party rule. "Any rejection of it would be tantamount to collective psychological suicide," Brzezinski wrote.

Yet the same forces that had prompted the attack on Brezhnevism and Stalinism would require erasing Leninism as well. The problem with all three isms, especially Leninism, was their common inheritance from grandfather Marxism: the belief, or predisposition to believe, that the economic and political development of large numbers of people could be entrusted to the virtues and abilities of a small elite. Hard Soviet—and Chinese—experience had proved this belief false. However well-intentioned an elite might be, it lacked sufficient wisdom for all the decisions that had to be made in a modern economy and society, and however well-intentioned the founding generation of an elite, declension invariably followed the passing of the founders. "In the final analysis," Brzezenski asserted, "Marxist-Leninist policies were derived from a basic misjudgment of history and from a fatal misconception of human nature." The end of the communist experiment was at hand. As a consequence, the end of the Cold War was at hand too.[23]

Francis Fukuyama took Brzezinski's theme of communism's failure and pushed it considerably further. Fukuyama, who worked for George Kennan's old outfit, the State Department's policy planning staff, suggested that the world was witnessing not merely the end of communism or the end of the Cold War, but the end of history. Curiously, considering his employer, Fukuyama's notion of history had little to do with the mundane business of relations among governments. Nor was Fukuyama's history chiefly a matter of wars or elections or migrations of people or capital, or any of the activities and issues that fill 90 percent of the pages of the world's history books. To Fukuyama, the history that counted was the history of ideas, and indeed a special subset of the history of ideas. At the core of history, he argued, was the struggle among intellectual paradigms for ordering relations among individuals. For nearly two hundred years, the leading paradigm had been Western liberalism, with its emphasis on personal and property rights. For the majority of those two centuries, history had been a matter of testing alternatives to liberalism. The most important alternatives had been communism and fascism. Fascism had lost out during World War II, when liberalism and communism combined to defeat it. After the war, at the beginning of the Cold War, liberalism and communism advanced to the finals.

For a time, communism gave liberalism a stiff challenge. The conversion of China to communism in 1949 enormously enhanced the credibility of the doctrines of Marx and Lenin, especially among the Third World peoples who were in the process of choosing between the liberal and communist paradigms. But communism ultimately failed on the crucial issue: economics. "The past fifteen years," Fukuyama wrote, "have seen an almost total discrediting of Marxism-Leninism as an economic system." From this discrediting, general disillusionment followed. In China, the government had allowed, even encouraged, a transition away from centralized control of the economy, and though China's rulers were trying to maintain their grip politically even as they devolved economically, Fukuyama believed they would encounter increasing difficulty doing so. In the Soviet Union, the retreat from communism had gone further in the political sphere than in the economic, but there, as in China, the loosening of communism's control in one area of national life would loosen it in other areas as well.

Fukuyama didn't claim that the Soviet Union and China would become liberal democracies on the Western model in the near future. This wasn't his point. His point was that the paradigm that once had energized Soviet and Chinese political and economic life was dead. "At the end of history it is not necessary that all societies become successful liberal societies, merely that they end their ideological pretensions of representing different and higher forms of human society." Communism's pretensions were dying fast. With the death, history ended.

Like others looking beyond the Cold War (and in his case, beyond history), Fukuyama viewed the future with something less than undiluted optimism. "The end of history will be a very sad time." Struggle was what gave life meaning, and the great struggles were nearly over. "The struggle for recognition, the willingness to risk one's life for a purely abstract goal, the worldwide ideological struggle that called forth daring, courage, imagination and idealism, will be replaced by economic calculation, the endless solving of technical problems, environmental concerns and the satisfaction of sophisticated consumer demands. In the post-historical period there will be neither art nor philosophy, just the perpetual caretaking of the museum of human history." Fukuyama confessed "a powerful nostalgia" for the time of struggle, and as a concluding obiter dictum he held out the possibility that "the prospect of centuries of boredom at the end of history" somehow would get history started again.[24]

The German question, and other chestnuts

Fukuyama needn't have worried about boredom. During the autumn of 1989, history hopped a fast train West, and within six months, the Cold War order in Europe vanished. The watershed event was the opening of the Berlin Wall in November, which not only undid the division of Europe, the most obvious manifestation of the Cold War, but also resurrected the German question—the issue that had triggered the Cold War in the first place. The desire to crush Nazi Germany had led to the initial partitioning of Hitler's Reich, and the failure of the wartime Allies to agree on what to do about postwar Germany had made the partition permanent— or at least long-lasting. The rearming of the West's portion of Germany, and Bonn's admission to NATO, had precipitated the creation of NATO's mirror image, the Warsaw Pact. This completed the conversion of what had begun as a diplomatic dispute between Washington (and London) and Moscow into the armed faceoff at Germany's heart that epitomized the Cold War.

As it became clear in the latter part of 1989 that the Brezhnev Doctrine for the Eastern European countries—once socialist, always socialist—had given way to what a Gorbachev spokesman called the Sinatra doctrine— they do it their way—American leaders and commentators remembered that the American presence in Europe, besides being designed to keep an eye on the Soviets, served the additional purpose of keeping an eye on the Germans. The historically minded recalled a statement by John Foster Dulles in 1954 explaining the caution with which the West should approach the subject of German (at the time, West German) sovereignty. Dulles declared that everyone, including the Germans, wished to avoid the danger of "resurgent militarism" in Germany. To prevent this, German rearmament must take place within the framework of a "collective international order." The collective order at hand was the North Atlantic alliance, although Dulles hoped for the evolution of a more comprehensive structure of European unity.[25]

During subsequent years, the European Economic Community came to complement the Atlantic alliance as an encouragement to German civility. But the Common Market lacked the crucial element required to keep the Germans satisfied with incomplete sovereignty: nuclear weapons. So long as the United States pledged to defend Germany, with American nuclear weapons if necessary, Bonn denied itself the big bombs and satisfied itself with the status of a second-rate power.

The arrangement worked well enough as long as the threat from the East persisted. Americans accepted the need for a strong commitment to Europe, and Germans accepted the need for American protection. As the Soviet threat diminished, though, at the end of the 1980s, the major parties to the arrangement began rethinking the deal. In the United States, would-be spenders of a post–Cold War "peace dividend" challenged the indispensability of stationing hundreds of thousands of American soldiers in Germany. Germans, who for a generation had put up with NATO maneuvers that trampled fields, clogged roads, rattled windows, and produced the impression that Germany was still an occupied country—which it was, albeit by friendly forces—looked toward the end of the Cold War as affording some peace and quiet.

Yet the sobering thought of an unattached Germany, especially one augmented by the reunification that appeared increasingly inevitable, worked to counteract the American inclination to pull back from Europe. The Cold War might be over, and the Soviet threat might be dissipating—although one could never be completely confident about a country that, economic ruin or not, possessed more than ten thousand nuclear weapons—but the system of international relations that had undergirded the Cold War still promised a measure of comfort for the uncertain years ahead. American officials refrained from voicing fears that Germany might again prove Europe's loose cannon. They left the voicing of such fears to less reticent types in the media and the think tanks (and to a British cabinet officer who got the boot for saying what many people on both sides of the Atlantic were thinking). As a consequence of this delicacy, Washington's demand that a unified Germany be a member of NATO seemed at times to be informed more by inertia than by logic. Logic there was, though, for whatever the defects of the Cold War, while it lasted one knew what to expect of much of the world from one day to the next. As the Cold War ended, each day brought fresh surprises. For the time being, most of the surprises were pleasant, for the West. But one could only guess how long the pleasantness would last. The American insistence on holding Germany within NATO represented an effort to keep the surprises within bounds.

The Soviets initially objected, although whether from wounded pride, sincere concern at a unified Germany in the enemy camp, or a desire to drive up the price of later concession, was hard to say. Eventually the Kremlin decided it might as well accept with reasonable grace what it couldn't reasonably prevent. The Soviet leadership may also have decided that a Germany under NATO's watch, and therefore not feeling pressured to develop such disconcerting emblems of independence as nuclear weapons, was preferable to a Germany all alone. Whatever the reasoning, in the summer of 1990, Moscow cut a deal with Bonn (already preparing to relocate to Berlin, everyone guessed) discreetly tying Soviet acquiescence in German NATO membership to German economic aid to the Soviet Union and some face-saving promises about not positioning troops from other NATO countries in the soon-to-be-former German Democratic Republic.

If the re-knotting of the Germanys convinced nearly everyone that the Cold War was over, the Persian Gulf crisis that began in August 1990 convinced the rest. By invading Kuwait, Saddam Hussein accomplished the heretofore nearly inconceivable feat of arraying both Americans and Soviets on the same side in a war-threatening situation in the Middle East. With rare and fleeting exceptions like the Suez crisis of 1956, Washington and Moscow previously had taken great pains to assume conflicting positions whenever the principal powder keg (and petroleum barrel) of the world threatened to explode. When the United States backed the Israelis, the Soviet Union backed the Arabs. When the United States cultivated Iran, the Soviet Union cultivated Iraq. When the United States wooed Egypt, the Soviet Union wooed Syria. When the United States supported Saudi Arabia, the Soviet Union supported Yemen. But following the Iraqi invasion of Kuwait, Washington and Moscow joined hands in condemning Saddam Hussein and guiding sanctions through the United Nations Security Council.

Only the very optimistic ascribed this diplomatic conjunction to a Russian conversion to America's kind of right-mindedness on the need for inviolable frontiers and the protection of the small and weak against the large and strong. For that matter, it required almost equal optimism to believe in the White House's complete conversion to such views. In fact,

Washington and Moscow arrived at the same destination in the Persian Gulf by different routes. The Americans wanted to keep Hussein's hand off the oil spigot of Kuwait and Saudi Arabia as least as much as they wanted to guarantee self-determination (of an autocratic sort) to the Kuwaiti people. The Soviets wanted to act the part of responsible world citizens in order to ensure a supply of Western capital to underwrite their leap to a market economy.

Yet whatever the origins of the conjunction, there it was. How long it would hold was another question. As an oil exporter, the Soviet Union had nothing against the higher prices Hussein was aiming for. As an oil importer, the United States did. As a longtime opponent of monarchy and Western imperialism, the Soviet Union had no philosophical interest in guaranteeing borders devised by Kuwaiti emirs and British colonialists. As a longtime (if not perfectly consistent) supporter of international legal forms, the United States did. At a couple of moments, just before the diplomatic crisis gave way to war in January 1991, and again before the anti-Iraq air offensive was succeeded by a ground invasion, Gorbachev started to waffle. Conservatives at home pressured him to put some distance between his government and the United States, and he did. But not much. In each case, the advantages to be gained by sticking with the majority of United Nations opinion seemed to outweigh the advantages of taking Iraq's point of view, and the Kremlin remained on the American side of the issue.

Past due

Though the anti-Iraq coalition held, Gorbachev's domestic position didn't. In August 1991, a group of disgruntled apparatchiks and worried military officers attempted to reverse five years of reform by arresting Gorbachev and seizing control of the Soviet government. The coup failed, almost farcically. Yet while Gorbachev regained his position as president of the Soviet Union, the net result of the affair was to shift the political center of gravity in the country to advocates of republican (that is, provincial) autonomy. Boris Yeltsin, president of the Russian republic, led a mass exodus from the Soviet Union. By the end of the year, the union was no more, replaced by a loose and seemingly tenuous Commonwealth of Independent States.

The dissolution of the Soviet Union convinced even those few who still doubted that the Cold War was over. If the Soviet government couldn't keep its own territory together, it didn't pose much of a threat to anyone else's. As the fissures began to develop in the edifice of Soviet power, some observers in the West feared a nuclear civil war between seceding Soviet republics. But nothing of the sort happened, and the creation of Lenin and Stalin went to its grave with a whimper, rather than a bang.

In the immediate aftermath of the opening of the Berlin Wall—the event, it looked increasingly clear, that future generations would recognize as marking the end of the Cold War—most Americans had been happy to celebrate a brilliant victory over their foe of forty years. In many ways, it was indeed a brilliant victory. The United States had gone head-to-head with the Soviet Union for nearly two generations, in Europe, Asia, Africa, and Latin America. The contest had been both geopolitical and ideological. The weapons of combat had been military, economic, and diplomatic. Sometimes the fight had been carried on in open view of the world, sometimes in the shadows of the international nether realm. At the end, the Soviet Union was utterly vanquished. What once had been a superpower lay shattered in more than a dozen pieces. Its motivating ideology was completely discredited, abandoned by all but a declining handful of stubborn Stalinists in China, North Korea, Vietnam, and Cuba.

Yet, as the buzz of watching Berliners dance atop the wall wore off, Americans increasingly asked themselves just what they had won. The end of the Cold War hadn't brought peace to the world, not even to Americans. On the first anniversary of the opening of the Berlin Wall, half a million American soldiers and sailors were in the Persian Gulf region or on their way there. Within two months more, they were engaged in the largest American military operation since before the Cold War began. The Persian Gulf War of 1991 didn't last long, and claimed relatively few American lives. For the first time since the 1940s, a Middle Eastern conflict hadn't threatened to drag in the superpowers on opposite sides. Even so, despite the thrashing Saddam Hussein received, the outcome of the war was disappointing. Hussein remained in power in Baghdad, as obstreperous, if not as dangerous, as ever. The Arab-Israeli dispute remained unresolved (although the parties were talking off and on). The industrialized nations remained dependent on the oil of the Persian Gulf. This last problem appeared likely to get worse, as the collapse of the Soviet Union turned the world's largest oil producer into a net importer. The successor states, along with the countries of Eastern Europe, soon would be competing on the world market for petroleum.

Nor had the end of the Cold War brought prosperity to the United States. On the contrary, the beginning of the 1990s witnessed a recession in America, the first in a decade. The timing was partly coincidental. All economic expansions run out of steam sooner or later. But the large cutbacks in weapons procurement consequent to the Cold War's end exacerbated the downturn, and in defense-oriented communities from New England to Southern California, people began to remember the bad old days fondly. Few were so blunt as to say that the Pentagon ought to keep buying weapons merely to provide jobs for defense workers (although candidate for re-election George Bush came close in 1992), but there was much talk of the need to proceed carefully—that is, slowly—in converting the economy to a post–Cold War footing.

For awhile, it had looked as though the Pentagon might not have to cut back much at all. Since the period when Gorbachev had commenced accepting American positions on arms control, the American military establishment had been on the lookout for ways to ensure that it would continue to be supported in the style to which it was accustomed. Perennial bears in the market of international relations, the generals and admirals and their civilian attachés found the bullish post–Cold War world environment distinctly unsettling. A measure of the Pentagon's concern was its agreement to enlist in America's war on drugs, which it previously had dismissed as distracting from the military's main mission. But drugs were small potatoes, and not even the Defense Department's best pitchmen could justify star wars and aircraft-carrier battle groups as necessary to neutralize the Medellin cocaine cartel.

Saddam Hussein arrived just in time, or so it seemed. More convincingly than the Pentagonists could ever have done on their own, Baghdad's bully demonstrated that there remained a role for American military power. Indeed, the Defense Department could hardly have devised a better demonstration. The campaign against Iraq allowed the Pentagon to show off its high-tech weaponry, including radar-evading Stealth aircraft, ground hugging cruise missiles, Patriot anti-missile missiles, and laser-guided bombs, as well as planes, helicopters, and tanks equipped with infra-red, fight-in-the-night viewers. Merely preparing for the war stretched America's sealift and airlift capacity to the maximum, which reminded Congress that the less sexy items of the defense budget mattered as much as the sleek and shiny.

An argument could even be made, and was, that the war in the Persian Gulf underlined the need for star wars. When technical studies had revealed—to the satisfaction of most, though not all, observers—the impracticality of building a shatterproof bubble that would protect the United States from a massive Soviet nuclear attack, SDI proponents suggested that a space shield was still worth building as a defense against stray shots from minor nuclear powers. At first, this was a tough sell, since the only minor nuclear powers were Britain, France, China, probably Israel, maybe India, and conceivably South Africa—none of which seemed motivated to go after the United States in the likely future. But Iraq was a potentially different story. Hussein didn't bother to disguise his desire for nuclear weapons, nor his disdain for much that Americans held dear. To be sure, although Baghdad appeared to be within a few years of developing usable nuclear warheads, no one expected Iraq's rocketeers to produce intercontinental missiles before the century's end. On the other hand, star wars wouldn't be ready much before then either. When America's high-tech weapons worked better in the six-week war against Iraq than most people had predicted, the case for star wars seemed stronger still.

In certain respects, however, American weapons worked too well. By destroying Hussein's capacity to make war, the Pentagon fought itself out of that part of its job. As the recession-aggravated federal deficit ballooned during the year after the Persian Gulf War, the Bush administration couldn't resist demands for substantial slicing in defense programs.

The demands to cut defense were one sign of the trepidation with which many Americans viewed the future at the end of the Cold War. A large part of the trepidation centered on the prospects for the American economy. Though the early and middle years of the Cold War had seen the economy surge forward, the last decade of the contest had witnessed a decline in America's economic health. The most obvious symptoms of the decline were the country's twin deficits: the federal budget deficit and the trade deficit. In the pre-Keynes era, a country's economy had customarily been considered much like the weather: everyone talked about it, but no one could do much about it. In the early 1990s, the federal deficit seemed to have fallen into a similar category. Politicians and editors expounded ceaselessly on the burden a large and chronic deficit would place on the shoulders of future generations of American taxpayers, with interest on the federal debt, now the largest single budget item, being the one that would have to be paid first. They decried the deleterious effects of the deficit on American international competitiveness, with safe government bonds drawing dollars away from risky but potentially rewarding, and productivity-raising, investments in the private sector. They wrung hands regarding even America's independence, with the world's largest debtor nation having to think carefully about actions that might frighten the foreign money required to roll over the federal debt. But despite all the talk, American politicians made precious little progress trimming the deficit, for the simple reason that cutting the deficit threatened to generate more politically mobilizable pain than politically mobilizable relief. The pain would hurt in the present, while the relief would ease matters only over the long term. Since 87 percent of American federal legislators (435 representatives and 33 or 34 senators, out of 535 total lawmakers) face election every two years, there is a strong bias in favor of the short term.

While the federal deficit indicated a failure of the American political system, the trade deficit reflected a more fundamental faltering of the American economy. At the beginning of the Cold War, American producers had been the most efficient in the world, and the free-trade regime the United States had sponsored worked significantly to their benefit. By the 1980s, however, American costs of manufacturing commonly exceeded those of foreign competitors. In one industry after another, Americans lost primacy to the Japanese, the Germans, or the newly industrializing countries of the Pacific rim. Whether America could run a large trade deficit indefinitely occasioned as much debate as how long the federal budget deficit could persist. Indeed, the twin deficits were Siamese, in that the trade deficit provided the foreign dollars that helped finance the budget deficit, while the budget deficit diverted investment capital that might have alleviated the trade deficit.

What no one could debate, however, was the fact that the two deficits revealed a considerable slippage of American power since the beginning of the Cold War. Though America's economy remained the world's largest, it was losing ground. If the European Community achieved effective economic integration as anticipated, the American economy would fall to second place for the first time during the twentieth century. Already the United States lacked the kind of discretionary wealth it had commanded early in the Cold War, when it could simultaneously fight the Korean War, rebuild the American military, and reconstruct Western Europe and Japan. Significantly, when the reform governments of Eastern Europe and the Soviet Union appealed for help in making the transition from command to market economies, it wasn't the Americans but the Europeans and Japanese who had the most cash on hand to help finance the switch. And when George Bush sent American troops, ships, and planes to the Persian Gulf, the American secretary of state, James Baker, toured the capitals of the world soliciting funds to finance the operation.

America's economic woes weren't entirely hangover effects from the Cold War. The budget-bingeing of the 1980s included growth in federal programs that had nothing to do with defense, programs that proved even more resistant than the Pentagon's pets to trimming. Moreover, anyone with historical perspective, viewing the world situation in 1945, would have declared America's decline relative to other countries to be inevitable. Eventually, those countries flattened by the world war—especially Germany and Japan—would pick themselves up and regain their places of prominence in the international community. In addition, industrializing countries—whether re-industrializing, like Germany and Japan in the 1950s, or newly industrializing, like Korea and Taiwan in the 1960s and 1970s—naturally tend to grow faster than already industrialized countries like the United States. Increases in productivity come more easily at low levels of production than at ones already high, and more readily in manufacturing than in the service activities that typify mature economies. The low-wage regimes in the newly industrializing countries worked to high-wage America's detriment as well.

Yet if a combination of factors produced the predicament Americans found their economy in at the beginning of the 1990s, the Cold War counted considerably in the combination. The military buildup of the 1980s had contributed significantly to the federal deficit, both in plain monetary terms and in the political sense of breaking government spending loose from the constraints of a broad feeling that budgets ought to balance, at least during prosperous times. By the mid-1980s, budget numbers that would have buried previous administrations were optimistic targets for would-be deficit-reducers. Moreover, the deficit became addictive. Economically, it fueled the expansion of the 1980s, and in the process entered into American expectations about the future direction of the economy, to the extent that mere talk of a balanced budget brought on withdrawal symptoms: that is, fears of recession. Politically, American officials and candidates for office got used to telling voters they could have what they wanted without having to pay for it. The voters responded by electing the candidates who promised most persuasively.

Meanwhile, the military emphasis of the American economy undermined American international competitiveness. For nearly two generations, the demands of the American military had channeled some of the most capable minds and personalities in the country into activities that did little for the long-term growth of the economy. That little wasn't nothing. The Manhattan Project and its Cold War successors had fostered the creation of the commercial atomic-power industry. Production for the American air force subsidized manufacturers of civilian aircraft. Various technologies crossed into the private sector from work done for the Pentagon and the space program (the latter itself a manifestation of the Cold War, beat-the-Russians mindset). But military-driven research and development yielded civilian-useful advances only inefficiently—and reluctantly, in that most of the cutting-edge work was originally secret. It wasn't coincidental that Germany and Japan, freed by the United States from the necessity to defend themselves by their own efforts, demonstrated particular aptitude at developing products for the more economy-expanding civilian market.

In the final accounting, a country can only support the foreign policy its economy can finance. At the beginning of the Cold War, the American economy could finance almost any kind of foreign policy the American people and government chose. The people and government chose an enormously ambitious policy, one that projected the United States vigorously into every time zone and every inhabited latitude. By the 1990s, the American economy could finance nothing so ambitious, partly because the economy itself lacked the resilience and dynamism it had possessed half a century earlier, partly because demands unrelated to foreign policy were claiming a larger share of the economy's output, and partly because other wealthy countries were bidding up the price of power. At the century's midpoint, $12 billion in Marshall Plan aid had rebuilt much of Western Europe. In the century's last decade, the same amount might have purchased a few blocks of downtown Tokyo. Even after adjusting for inflation and the overheated character of the Japanese real-estate market, the difference in purchasing power was enormous. The result was that United States was reduced from being head, shoulders, and torso above the rest of the world to being perhaps head above the rest. And some bumptious countries among that rest were growing fast, while the United States wasn't.

The passing of an age

The ambivalence that characterized much American thinking at the end of the Cold War reflected more than the troubled economy. It indicated a sense that Americans were witnessing the end of an era, an era that in many respects had been a golden age for the United States. The country had never experienced such a period of prosperity. With but a few glitches, the years of the Cold War had seen Americans grow wealthier and wealthier. The American standard of living at the Cold War's end was higher than it had ever been, and if pockets of poverty persisted, and if the country's growth rate had tailed off toward the present, nothing in life is perfect. Nor had America ever been more powerful in world affairs than during the Cold War. Indeed, no country in history had ever been more powerful than the United States had been, especially during the Cold War's early phase. It can make a nation giddy to bestride the world, as America had done from the mid-1940s to the 1960s. Heightening the giddiness was a recognition of the apparent irresistibility of American culture, which persisted into the 1990s. On every continent, in nearly every society, wherever one looked, people were emulating American styles of dress, watching American television shows and movies, eating and drinking American foods and beverages. At least outwardly, the world was becoming more like America all the time.

Yet perhaps the most important feature of the Cold War era, and that which would be missed most, was its conceptual simplicity. Charles Krauthammer was probably right when he argued that nations need enemies. During the Cold War, America had an enemy that could hardly have been improved upon. The Soviet Union was officially atheistic, which earned it the hostility of America's semi-official Christian majority. It was dictatorial, which offended American democratic sensibilities. It was socialistic, which threatened the private-property rights most Americans enjoyed or aspired to. It was militarily powerful, which endangered America's physical security. It was ideologically universalist, which set it in direct opposition to the United States, which was too. It was obsessively secretive, which precluded knowing just how dangerous it was. Should interest or inclination inspire one to inflate the threat, or simply to err on the side of safety, disproof or convincing correction was almost impossible.

While the Soviet Union remained a credible foe, Americans could congratulate themselves on their own relative goodness. Only the most morally chauvinistic thought America had a corner on the world's supply of *absolute* goodness, but so long as the Kremlin played its malevolent part, Americans merely had to be better than their Soviet rivals to feel virtuous. If America's record on race relations contained flaws, at least those who protested weren't packed off to gulags. If the American political system was sometimes superficial, at least Americans got to vote in genuine elections. If America occasionally settled for less than democratic perfection from its allies and clients, at least it held up a democratic example for them to follow.

Beyond the realm of moral psychology, the Cold War framework simplified the problem of understanding international relations. The bipolar scheme of world affairs reduced the need to delve closely into the motives and objectives of other countries, since a country's position regarding communism served, in the predominant American view, as a litmus test for that country's policies as a whole. To be sure, this test sometimes yielded false results. Neutralists like India often found themselves treated as fellow-travelers, while some right-wing dictators were accorded Free World membership. Moreover, those American officials who made their careers in foreign policy never took the litmus results very seriously. But for Americans who desired a quick and dirty division of the world into friends and foes, the communism-versus-democracy test did the job.

The communist issue served a similar purpose in American domestic politics, although here the problem of spurious results was even worse. Few American voters cared to take the time to educate themselves to the nuances of the possible positions candidates might adopt on issues relating to national security. For that majority who didn't, the question of whether an individual was reassuringly hard or suspiciously soft on communism simplified the sorting process. The trouble was that once candidates caught on to the game, almost everyone passed the test.

The Cold War simplified matters for particular groups in other ways. The forty years of the Cold War were a glorious time for the American defense industry, which might have been accused of colluding with the Soviet defense industry had the latter enjoyed any ability to collude across borders. Collusion or not, the armorers of the two sides shared an interest in heavy defense spending, and they benefited from each other's arms-racing actions. In the absence of such a readily identifiable and consistently threatening enemy as the Soviet Union, American weapons-producers never would have achieved the growth they did during the post-1945 period. Their vested interest in the Cold War appeared in their bottom lines.

Persons and organizations that had hidden behind the Cold War to oppose social reform—regarding race, for example—had a less compelling interest in keeping the chill on East-West relations. For them, the Cold War had been a handy distractive device, but should it be taken away, they would find another—secular humanism, perhaps. All the same, red-baiting would be hard to match for its capacity to change the subject, and to throw advocates of reform on the defensive.

Just as the Cold War had simplified matters for many in the United States, its end promised to complicate things. Psychologically, Americans would have to adjust to a world lacking an agreed-upon focus of evil against which they could favorably contrast themselves. Possible alternative focuses fell short in one respect or another. Saddam Hussein served the purpose for awhile, but he didn't have the staying power required of a real solution to the problem. Japan took some heavy beating on the issue of trade and jobs, but even many Americans thought their country's economic woes were chiefly homegrown. Neither was as soul-satisfying as the Soviets.

Politically, American candidates and public officials would have to come up with a more imaginative national-security agenda than reactive anti-communism. Bush promoted his "new world order" for a time, but acceptance levels were disappointing. Japan-bashers in Congress ran up against the difficulty that Americans liked the goods Nissan and Sony were sending east.

Economically, advocates of high defense spending would have to devise new rationales for keeping the production lines humming. Hussein helped, but not for long. And in the absence of a new threat, the peace-dividenders likely would slash deeply into profits and jobs.

Strategically, American planners would have to figure out how to deal with a world unlike that which they had come to know over forty years. While the demise of the Soviet Union diminished the likelihood of a civilization-shattering thermonuclear war, other sources of tension quickly ruled out a nail-biting moratorium. The Persian Gulf crisis and war demonstrated that troubles would persist into the post–Cold War era. Strikingly, the fact that Washington and Moscow were cooperating in the affair, a fact often cited as evidence of the new possibilities for peace opened up by the Cold War's end, actually deprived Washington of a potentially important diplomatic lever. In previous regional conflicts, when the United States and the Soviet Union had backed opposing parties, American leaders could pressure Moscow to restrain its allies, in the interest of preserving or improving broader superpower relations. The pressure didn't always work as well as Washington hoped, as the Nixon administration's efforts to get Moscow to help stop the Vietnam War short of a North Vietnamese victory demonstrated. Yet even in the Vietnamese case, Hanoi accepted a ceasefire. And, generally, a client's desire for continued Soviet aid acted to moderate its behavior somewhat. In cases where the two superpowers took the same side, the Kremlin card lost its value.

In Europe, the Soviet withdrawal from the center of the continent, accomplished in principle if not in detail during the summer of 1990, rendered the American guarantee of German security almost worthless. The Germans were too business-minded to go on long paying something for nothing. This implied major changes in the structure of European and Atlantic relations. German reunification per se had little to do with the issue, if only because for the near future the annexation of East Germany by West Germany would cost rather than benefit the Germans. Nor did one have to suppose another nasty turn by German nationalism to predict that sooner or later Germany would begin to act more independently. Perhaps the European Community would provide the supra-national framework the Cold War previously had. Perhaps not. The latter possibility was what had Germany-watchers worried. Germany's insistence on swift recognition of the independence of Slovenia and Croatia during Yugoslavia's civil war, while the other European governments and the United States urged caution, didn't lessen the worries.

Future American relations with Japan raised similar concerns. Like the Germans, the Japanese had been persuaded to devote their considerable energies and ingenuity to perfecting the performance of their economy. Matters of defense and foreign affairs they left primarily to the Americans. Although a few nationalist-minded Japanese, and a somewhat larger number of cost-conscious Americans, had complained at the arrangement, it substantially satisfied both parties so long as Cold War Russia presented a plausible danger to Japanese and American interests. As the Soviet threat diminished, however, many Japanese grew less inclined to follow America's lead internationally, and many Americans grew less inclined to pay for Japan's defense. Analogously with the German case, one didn't have to assume a return to the militarism of the 1930s to wonder about the effects of the re-emergence of Japan as an independent East Asian great power. A trade war between the United States and Japan, the first rumblings of which already were echoing through Congress, would be bad enough.

Was Hitler normal?

The sudden end of the Cold War, succeeded in short order by the collapse of the Soviet Union, raised some fundamental questions regarding what it had been all about. The most obvious question was whether it had been necessary for Americans to get so worked up over an enemy that proved to be a shell—a large country, to be sure, with formidable-looking weapons, but one with a decrepit economy and a political will insufficient to keep it from breaking apart at the first wind of honest reform. Had the Soviet threat *ever* been very great? How much of the perceived threat had been genuine and how much a figment of American imaginations? Was "cold war" useful, or misleading, as a description of the rivalry between the United States and the Soviet Union?

The "cold war" metaphor had first gained general currency with Walter Lippmann's 1946 book, *The Cold War.* At the time Lippmann wrote, the metaphor was plausible enough. Americans had just fought the biggest war in history, and found themselves confronting circumstances that resembled, in certain respects, those that had preceded the war. Stalin was as much a dictator as Hitler had been, and the Red Army was indisputably powerful. Communist ideology was potentially as expansionistic as that of the Nazis, if less explicitly bellicose. Undeniably, Stalin was a person to keep an eye on.

But the cold-war metaphor worked better as a literary device than as a description of international reality. Wars, at least as Americans historically have understood and fought them, are relatively brief affairs, with readily distinguished enemies and concrete objectives. Sometimes Americans had gained their wartime objectives: independence from Britain, subjugation of the Confederacy, destruction of Hitler. Sometimes they hadn't: acquisition of Canada, preservation of secession and slavery. But in every instance, the enemies and the goals had been clear, and Americans could tell whether they had attained the goals or not.

The objectives of the Cold War were considerably more nebulous, as was the nature of the enemy. Was the enemy communism? Or was it the Soviet Union? Was China an enemy? Then how could it become a friend? Was the United States fighting for territory, or for political and moral principles? Was containment sufficient to America's needs, or must the United States roll back communism?

Had Americans been less beguiled by the Cold War metaphor—had it not served so many purposes beyond the realm of foreign affairs—they might have recognized that the Cold War was no war at all, but simply the management of national interests in a world of competing powers. Because Americans defined their interests globally, and because America's foremost rival possessed mighty military weapons, American interest-management involved incessant effort and careful weighing of the possibility of armed conflict. Yet, though it was new to Americans, this was the sort of thing great powers had done as long as there had been great powers. It wasn't the comparatively placid and uneventful peace Americans had gotten used to in their many years of relative insulation from world affairs, but neither was it war.

Whatever the validity of the Cold War metaphor, Americans during much of the post-1945 period operated according to the premise that the only way to prevent the Cold War from flaring into World War III was to prevent a replay of the events that had led to World War II. This premise rested on a second, more basic premise: that Hitler wasn't an aberration but an archetype, that the model of escalating aggression he had used would be used by other ruthless, ambitious, and powerful national leaders. As it pertained to Stalin and the Soviet Union, this premise was made explicit in the "red fascist" imagery of the early Cold War. But essentially the same idea at various times infused American thinking about such countries as China, North Korea, Vietnam, Cuba, Indonesia, Egypt, and Iraq. The thrust of the argument was that dictators are insatiable, that aggression feeds on weakness, that appeasement merely postpones the day of reckoning.

Without doubt, power creates a certain community among those who wield it. And those who employ force as their primary instrument of policy tend to respond more readily to counterforce than to less direct kinds of appeals. Even so, the Hitler analogy obscured at least as much as it illuminated. Prior to the outbreak of World War II, Stalin showed none of the territorially expansionist compulsions that made Hitler Hitler. If anything, Stalin's reign produced a retreat from the world-revolutionism of Lenin's era, and the Georgian strongman's chief contribution to Marxist-Leninist theory was the notion of "socialism in one country." Neither did Stalin's actions after the war demonstrate much beyond a stubborn desire to prevent a repeat of the recent ruination. The Red Army refused to withdraw from where the war's end found it, but the Kremlin captured no new territory by force. The only significant Soviet military actions after 1945 were the crushings of reform in Hungary in 1956 and Czechoslovakia in 1968, and the Afghanistan war of 1979 and later. The Hungarian and Czech operations, though brutal and morally repugnant, were plainly designed to bolster a tottering status quo, to hold what Moscow had, rather than to extend the Russian writ to fresh territory. The Afghanistan fighting was largely defensive as well, intended to ward off the advance of Islamic fundamentalism toward the Soviet Union's Muslim provinces.

For forty years, the United States and its NATO allies devoted tremendous effort to preparations for the defense of Western Europe against a Soviet attack. During all that time, the attack never came. Why not? Did the Kremlin decide, in the face of the Western preparations, to forget about adding West Germany or France to its European empire? Or had it never intended such additions?

There is no way of knowing. Stalin, like many dictators, took his secrets with him to his tomb. Certainly, the Soviet military had contingency plans for an attack against the West, but planning is what planners get paid to do, and many plans have almost nothing to do with reality. (Until the 1930s, American strategists were drafting contingency plans for a war against Britain.) Besides, attacks can be defensive. If you are convinced that the enemy is going to hit you, you'll probably want to hit first. While it is impossible to prove that Stalin did *not* intend to attack the West, neither has it been shown that he *did*.

In the early years of the post-1945 period, when the memories of World War II's horrendousness were still raw, when the Western European countries were in a comparatively exposed position, and when American resources almost overmatched the rest of the world combined, American leaders understandably preferred to err on the side of caution. Less understandably, they continued to err long after circumstances had changed. By the 1960s or 1970s or 1980s, one might have thought, the burden of proof should have been on the alarmists. But by then, of course, the Cold War had been thoroughly domesticated and bureaucratized, providing benefits only barely involving American national security.

If the Hitler analogy obscured what Stalin and the Soviets were up to, it made a mash of what other communists were about. Tito's nose-thumbing at Stalin should have demonstrated that communists could be as fractious among themselves as in relations with capitalists. And anyone with the least sense of Chinese history, or the slightest understanding of the traditional Chinese disdain for most things foreign, could have guessed that the Chinese would follow Moscow's line exactly as long as they discerned advantages to themselves from doing so.

But because it served other purposes—political, economic, psychological—to treat communism as a global conspiracy, and to liken a failure to confront this conspiracy to the failure to halt Hitler, anti-appeasement became the touchstone of American Cold War policy. In anti-appeasement's name, Americans fought a bloody war in Korea, believing they were frustrating the Kremlin's planet-devouring designs. While the Korean War yielded mixed results for South Korea—the fighting devastated the country, but left it beyond Kim Il Sung's obnoxious reach—the cost far exceeded anything Americans would have accepted simply for South Korea's sake. In anti-appeasement's name, the United States fought another bloody war in Vietnam, failing this time to achieve even the preservation of a noncommunist government. As if to drive home the lesson that fighting in Vietnam to contain China had been wrongheaded, communist China and communist Vietnam soon fell out, to the point of war in 1979.

Comparisons to Hitler had the perverse effect of overblowing the communist military threat, which was largely nonexistent, and consequently understating the communist political threat, which wasn't. To their credit, some American officials in fact understood that the threat the communists posed was principally political. To their discredit, they did relatively little to share their insight with the wider American public. The Cold War climate in America wasn't conducive to nuance regarding communism. A few brave souls tried to explain such matters as that a communist China needn't be a China unalterably wedded to Moscow and irretrievably antagonistic to the United States, but the personal calumny and professional banishment these few suffered for their efforts alerted their colleagues and successors to the consequences of divergence from the party line. Most American officials chose the safer route of looking on communism as a hungry beast poised to devour the world as soon as America's guard let down. Measures designed to counter the communist political threat—for instance, by improving the economic and political performance, and thereby the attractiveness, of the countries of the "Free World"—were occasionally enacted, with the Marshall Plan being the outstanding example. But for every dollar Washington spent on economic aid, and for every meaningful exhortation American officials made to allies to respect democratic rights, Washington spent a hundred dollars on weapons, and American officials gave a score of speeches calling for staunch resistance to communist aggression.

Most perversely, the call to arms against communism caused American leaders to subvert the principles that constituted their country's best argument against communism. In 1945, the United States stood higher in the estimation of humanity than ever before, arguably higher than any country had ever stood in history. American soldiers and sailors had played a central role in the recent defeat of the almost universally detested fascists. Unlike the British and French, the Americans had no extensive colonial holdings that gave the lie to their professions of support for self-determination—the Philippines were slated for (and received) independence in 1946, and Puerto Rico didn't appear to want it.

Unlike the Russians, the Americans treated the peoples of the countries they liberated with respect, and rather than seeming scarcely an improvement over the Wehrmacht, as Red Army troops often did, American GIs brought hope of an end to strife and oppression.

Within a short time, however, world opinion of the United States began to slide. The better to contain communism, Washington aligned itself with colonial and reactionary regimes that flouted the principles Americans had just fought a war to vindicate. Since comparatively few persons outside the North Atlantic region considered communism a greater enemy than colonialism and institutionalized inequality, what appeared a necessary tradeoff to many Americans appeared self-serving and hypocritical to most foreign observers of American actions. People of the Third World—which earned its sobriquet precisely because of its inhabitants' determination to resist the two-worlds framework of the Cold War—often deemed America's alliance-building tantamount to imperialism, of which they had had more than enough. American intervention in the Korean War looked to be misguided meddling in an Asian civil conflict. The war in Vietnam was widely viewed as a case of neo-colonial repression of indigenous nationalism. American backing for rightist regimes elsewhere in Asia, Africa, and Latin America seemed to fit the imperialist pattern.

Self-described "realists" in the United States could ignore the Third World carping—How many divisions did Nehru have?—and contend that any illusions anyone harbored of meaningful American moral superiority were better off debunked. The Cold War, they held, like all great-power conflicts, was essentially amoral. The strong did what they wanted, the weak what they were required, and there was little of right or wrong about the matter. This was the Doolittle philosophy: beat the devil at his own game. And it was the philosophy that, slightly disguised, informed the Kirkpatrick doctrine of support for right-wing dictators and antagonism to left-wing dictators.

The flaw in this philosophy was that it didn't suit the American people. Whatever the objective merits of "realism" as a description of behavior among nations, and whatever its appeal or lack thereof to Germans, Brazilians, Chinese, Nigerians, or anyone else, Americans have from the beginning of their national existence demonstrated an incurable desire to make the world a better place. Sometimes they settled for the stand-offish exemplarism of John Quincy Adams. Sometimes they insisted on the missionary interventionism of Woodrow Wilson. But almost always they believed that America had important lessons to teach their fellow human beings: about democracy, about capitalism, about respect for individual rights and personal opportunity and the rule of law.

This save-the-world inclination was largely responsible for the fervor with which Americans waged the Cold War. It provided much of the impetus behind the Marshall Plan: Americans were going to rescue Western Europe from starvation, disease, and despair. It lay beneath the commitment of American lives to Korea and Vietnam: the United States would preserve those vulnerable countries from the depredations of dictatorship. It motivated the appropriation of billions of dollars in aid to countries of Asia, Africa, and Latin America: American funding would help bring prosperity and dignity to the downtrodden of the planet. To be sure, the rhetoric of American concern for the welfare of other peoples and countries usually involved some hypocrisy, and in back of every important Cold War initiative there lurked careful considerations of self-interest. Yet the very fact that self-interest had to be dressed up in selfless clothing testified to the importance of moral factors in American politics, and consequently in American foreign relations. The staying power of the Cold War paradigm resulted in no small part from its capacity to combine the selfless with the self-interested.

Sometimes the twain parted, though, and when the parting became undeniable, as during the Vietnam-Watergate era, it rent the American Cold War consensus. All but the most hard-bitten Americans found sorely trying the discovery that the United States government had actively sought to assassinate leaders of foreign countries, countries not at war with the United States, whose principal crime consisted of being caught in the crossfire between the White House and the Kremlin. Americans of every political persuasion recoiled from the televised images of South Vietnamese citizens immolating themselves to protest the policies of the government the United States was supporting, from the photographs of naked Vietnamese children running screaming from napalm attacks by American planes, from the descriptions of Vietnamese villages destroyed that they might be "saved." Liberals and conservatives alike resented the corruption of the American political and legal process in the name of national security.

Defenders of American Cold War policies held that compromises were necessary to defend basic American values. Whether they really *were* necessary is impossible to tell. Conceivably, had American agents not conspired in the overthrow of popularly based governments in Iran and Guatemala; had they not tried to assassinate Lumumba and Castro; had they not tampered with elections in the Philippines and Syria and elsewhere; had they not destabilized leftist regimes in Chile and other countries; had they not bombed Indonesian islands and mined Nicaraguan harbors; had the United States not provided arms and money to a score of repressive juntas from Cuba to Pakistan to Zaire; had the FBI not disrupted the lawful activities of legitimate political groups in the United States; had the CIA not violated its own charter and engaged in domestic espionage; had American armed forces not lost 50,000 dead in Korea and nearly 60,000 in Vietnam—conceivably, had the United States not committed these acts, along with other acts presumably more constructive, communism might have conquered the world, or enough of it to render America significantly poorer, unhappier, and less secure.

In the real world, however, what counts isn't the conceivable but the likely. From the vantage point of the 1990s—which, of course, isn't the vantage point of the late 1940s and 1950s—the internal weaknesses of communism seem to have been sufficiently great to have made anything approaching the world-conquest scenarios of NSC 68 and similar manifestoes exceedingly improbable. Although conservatives claimed that American pressure was responsible for finally buckling the Soviet system in the late 1980s, as reasonable a case can be made that American antagonism actually *prolonged* the Cold War. For almost forty years, while Soviet leaders could plausibly cite an American threat to the security of their country—and, considering Washington's success in ringing Russia with American allies, considering the large and ever-growing size of the American nuclear arsenal, and considering the "massive-retaliation" and "evil-empire" language American leaders recurrently resorted to, the threat must have seemed quite plausible—Moscow could put off dealing with the problems inherent in the communist scheme of government. Had the United States not cooperated in playing the villain (just as the Soviet Union played the villain to the United States), the Kremlin might have been forced to confront its true problems sooner.

Similar considerations apply to America's dealings with other communist countries. By backing Chiang in China's civil war, for nearly a generation after that war otherwise would have ended, the United States handed Beijing's new mandarins an issue with which to divert the Chinese masses from their overwhelming domestic difficulties. Washington was consistently Fidel Castro's best friend by being his worst enemy. If Castro had ever had to justify his one-man rule in Cuba on its own merits, rather than on the demerits of the superpower across the Florida Strait, whose leaders still tried to strangle the Cuban economy thirty years after the Cuban revolution, he would have found the going a great deal harder. A principal consequence of American involvement in the Vietnamese war—aside from the millions of deaths, maimings, and displacements the fighting in Indochina produced—would seem to have been the postponement of the day when the Vietnamese communists had to stop fighting, at which they were very good, and start governing, at which they were a disaster.

The fact is that communism—not capitalism or democracy—has been the communists' worst enemy. But nations have had to discover this for themselves. External force has usually succeeded only in delaying the discovery.

Had attempts to force the discovery been costless, the delay might not have meant much. But the cost to America, not to mention to the delayed nations, was very high. More than 100,000 Americans died fighting wars that had almost nothing to do with genuine American security. The American economy, in 1945 the envy of the earth and the engine of global growth unprecedented in history, by the 1990s sputtered and faltered under the weight of four decades of military spending inconceivable before the Cold War. The chronic deficits that were a primary legacy of that military spending prevented the federal government from addressing many of the serious problems that crowded in on the country. Perhaps worst of all, American leaders, sometimes without the knowledge of the American people, sometimes with the people's approval, consistently cut moral corners in the Cold War, contradicting the ideals America was supposed to be defending. In 1945, nearly all Americans and probably a majority of interested foreigners had looked on the United States as a beacon shining the way to a better future for humanity, one in which ideals mattered more than tanks. During the next forty years, American leaders succeeded in convincing many Americans and all but a few foreigners that the United States could be counted on to act pretty much as great powers always have. If Americans felt ambivalent about their victory over the Soviet Union, they had reason to.

Notes

Chapter 1 The Last Days of American Internationalism: 1945–1950

1. Yalta conference communique, 2/12/45, *Foreign Relations of the United States 1945*, Malta and Yalta conferences: 970–71.
2. Bohlen minutes of dinner meeting, 2/4/45, ibid., 589.
3. Ibid.; Bohlen minutes of second plenary meeting, 2/5/45, ibid., 617.
4. Matthews minutes of third plenary meeting, 2/6/45, ibid., 677–80.
5. Roosevelt to Stalin, 2/6/45, ibid., 727–28.
6. Yalta conference communique, 2/12/45, ibid., 973.
7. Acheson testimony, 11/30/44, U.S. Congress, House of Representatives (78th Congress, 2nd session), *Hearings Before the Special Committee on Post-war Economic Policy and Planning of the House of Representatives* (Washington, 1945), 4:1081–83.
8. Ibid., 1074–78.
9. Dean Acheson, *Present at the Creation* (New York, 1969), 133.
10. Report of the ad hoc committee of the State-War-Navy Coordinating Committee, 4/21/47, *Foreign Relations 1947*, 3:210–11.
11. Clayton memo, 5/27/47, ibid., 230–32.
12. Marshall address, 6/5/47, *Department of State Bulletin*, 6/15/47.
13. Eisenhower to Patterson and Forrestal, 3/13/47, *Foreign Relations 1947*, 5:110–14.
14. Porter to Clayton, 2/17/47, ibid., 17–22.
15. State-War-Navy Coordinating Committee report, [3/3/47], ibid., 76–78.
16. Truman address to Congress, 3/12/47, *Bulletin*, 3/23/47.
17. Herbert L. Matthews, "Fascism Is Not Dead," *Nation's Business*, December 1946.
18. Washington and Jefferson quoted in Norman A. Graebner, ed., *Ideas and Diplomacy* (New York, 1964), 75, 78.
19. *Congressional Record*, 7/11/49.
20. Ibid., 9209.
21. MacArthur in *New York Times*, 3/2/49.
22. Acheson, *Present at the Creation*, 450.

Chapter 2 The National Insecurity State: 1950–1955

1. NSC 68, 4/7/50, *Foreign Relations 1950*, 1:235–92.
2. Acheson, *Present at the Creation*, 374–75.
3. NSC 68.
4. Niebuhr, "Why Is Communism So Evil?," *New Leader*, 6/8/53, reprinted in *Christian Realism and Political Problems* (New York, 1953).
5. Statistics in this section are based on U.S. Department of Commerce, *Historical Statistics of the United States* (Washington, 1975).
6. Eisenhower speech, 4/16/53, *Public Papers of the Presidents: Dwight D. Eisenhower 1953* (Washington, 1960).
7. Jackson to Rockefeller, 11/10/55, C. D. Jackson papers, Eisenhower Library, Abilene, Kansas.
8. Task Force A summary, undated (enclosed in memo by Lay, 7/22/53), *Foreign Relations 1952–1954*, 2:399–412.
9. Task Force B summary, ibid., 412–16.
10. Task Force C summary, ibid., 416–31; Group C summary, 7/16/53, records of the Office of the Special Assistant for National Security Affairs, Eisenhower Library.
11. NSC 162/2, 10/30/3, *Foreign Relations 1952–1954*, 2:577–97.
12. Dulles address, 1/12/54, *Bulletin*, 1/25/54.
13. Robert Bowie oral history, Princeton University.
14. Memo of NSC meeting, 3/10/55, *Foreign Relations 1955–57*, 2:345–50; Radford to Wilson, 3/27/55, 091 China, Radford files, Joint Chiefs of Staff records, National Archives, Washington.
15. Memo of NSC meeting, 3/10/55, *Foreign Relations 1955–57*, 2:345–50; memo of conversation, 3/26/55, ibid., 2:400.
16. House of Representatives, Committee on Appropriations (83:2), *Department of the Army Appropriations for 1955* (Washington, 1954), 43, 49.
17. Matthew B. Ridgway, *Soldier* (New York, 1956), 272–73, 327.
18. Taft speech, 2/8/51, *Congressional Record*.
19. Mohamed Hassanein Heikal, *The Cairo Documents* (Garden City, N.Y., 1973), 40; Heikal, *Cutting the Lion's Tail* (New York, 1987), 39.

Chapter 3 The Immoral Equivalent of War: 1955–1962

1. Donovan to Roosevelt, 11/18/44, Allen Dulles papers, Princeton University, Princeton, New Jersey.
2. National Security Act of 1947 (P.L. 253, 7/26/47); NSC 10/2, 6/18/48, reprinted in William M. Leary, ed., *The Central Intelligence Agency: History and Documents* (University, Ala., 1984).
3. Background briefing, 4/3/53, Allen Dulles papers.
4. Memo of conversation, 10/19/54, Eisenhower papers, Eisenhower Library.
5. Doolittle report quoted in United States Senate, Select Committee to Study Government Operations with Respect to Intelligence Activities (94:2), *Supplementary Detailed Staff Reports on Foreign and Military Intelligence* (Washington, 1976), 4:52–53, note 9.
6. King to Dulles, 12/11/59, quoted in United States Senate, Select Commit-

tee to Study Government Operations with Respect to Intelligence Activities (94:1), *Alleged Assassination Plots Involving Foreign Leaders* (Washington, 1975), 92.

7. Memo for record, 3/9/60, ibid., 93.

8. Minutes of meetings, 7/21/60 and 8/18/60, ibid., 57–58.

9. Dulles to Leopoldville, 8/26/60, ibid., 15.

10. Leopoldville to Dulles, 9/20/60, ibid., 18.

11. Joseph Scheider testimony, 10/7/75, ibid., 20–21.

12. Leopoldville to Tweedy, 11/14/60, ibid., 33.

13. Elisabethville to Dulles, 1/19/61, ibid., 51.

14. Killian report, 2/14/55, Office of the Special Assistant for National Security Affairs records, Eisenhower Library; memo of discussion of NSC meeting, 9/18/55 (memo dated 9/15/55), Eisenhower papers, Eisenhower Library; JCS 1899/53, 8/8/53, 381 U.S. (5–23–46), Joint Chiefs of Staff records, National Archives.

15. "Deterrence and Survival in the Nuclear Age" (Gaither report), 11/7/57, published for the Joint Committee on Defense Production, 94th Congress, 2nd session (Washington, 1976).

16. *New York Times*, 11/15/57.

17. Eisenhower to Cowles, 10/27/53, Eisenhower papers.

18. Diary entry for 9/16/47, Robert H. Ferrell, ed., *The Eisenhower Diaries* (New York, 1981), 143.

19. Eisenhower, *Waging Peace* (Garden City, N.Y., 1965), 624–625.

20. Malik oral history, Princeton University; Dulles to Jackson, 8/24/54, C. D. Jackson papers, Eisenhower Library.

21. Dulles memo, 6/29/55, Dulles papers, Eisenhower Library.

22. Edward Teller and Albert L. Latter, *Our Nuclear Future* (New York, 1958), 136.

23. Kahn quoted in Paul Dickson, *Think Tanks* (New York, 1971), 109.

24. Rickover testimony in United States Senate (90:2), *Hearings before the Committee on Foreign Relations*, 5/28/68, pp. 12, 22.

25. Allen Dulles briefing of newspaper correspondents, 4/3/53, Allen Dulles papers, Princeton University.

26. Eisenhower news conference, 5/11/60, *Public Papers*.

27. George B. Kistiakowsky, *A Scientist at the White House* (Cambridge, Mass., 1976), 375.

28. Memo by McManus, 1/19/62; testimony by unnamed executive assistant, 6/18/75, *Alleged Assassination Plots*, 334–36.

29. LeMay quoted by Robert McNamara in *New York Times Magazine*, 8/30/87.

30. Minutes of meeting, 10/16/62, in *International Security*, Summer 1985; minutes of meeting, 10/27/62, ibid., Winter 1987–88.

31. Minutes of meeting, 10/27/62, ibid.

32. *New York Times Magazine*, 8/30/87; George H. Gallup, *The Gallup Poll* (New York, 1972), 3:1786–93; Theodore Sorensen, *Kennedy* (New York, 1965), 705; Arthur Schlesinger, Jr., *A Thousand Days* (Boston, 1965), 841.

Chapter 4 The Wages of Hubris: 1962–1968

1. *Congressional Record*, 7/2/57.

2. William Appleman Williams, *The Tragedy of American Diplomacy* (New York, 1959, 1962), 305–7.

3. C. Wright Mills, *The Power Elite* (New York, 1956, 1959), 205–6.

4. Paul A. Baran and Paul M. Sweezy, *Monopoly Capital* (New York, 1966), 187, 209.

5. Robert Welch, *The Politician* (Belmont, Mass., 1964 ed.), 276–77, 300.

6. William F. Buckley, Jr., "The Uproar," *National Review*, 4/22/61.

7. Barry Goldwater, *The Conscience of a Conservative* (Shepherdsville, Ky., 1960), 87–90, 94, 111–12, 118–23.

8. Bundy quoted in Melvin Small, *Johnson, Nixon and the Doves* (New Brunswick, N.J., 1988), 33.

9. Notes of Johnson meeting with Bob Thompson, 8/21/67, Meeting notes file, Johnson papers, Johnson Library, Austin, Texas.

10. Small, *Johnson, Nixon and the Doves*, 78.

11. Hoover memos, 4/28/65, excerpted in U.S. Senate, *Final Report of the Select Committee to Study Government Operation with Respect to Intelligence Activities* (Washington, 1976), 2:251.

12. Hoover memo, 4/28/65, ibid.

13. FBI memo attached to Hoover to Bundy, 4/28/65, ibid.

14. Taylor Branch, *Parting the Waters: America in the King Years 1954–63* (New York, 1988), 211.

15. Hoover memo, 10/12/61, reproduced in Cathy Perkus, ed., *COINTELPRO: The FBI's Secret War on Political Freedom* (New York, 1975), 19.

16. FBI Detroit office to Hoover, 11/2/61, ibid., 44–45.

17. FBI Denver office to Hoover, 5/4/65, ibid., 50–51; FBI New York office to Hoover, date not given, ibid., 52–53; Hoover to New York, 10/8/69, ibid., 54.

18. CIA headquarters instructions, August 1967, excerpted in *Final Report*, 2:100; Helms to Kissinger, 2/18/69, ibid., 101.

19. *Columbia Record*, 5/5/64; *Charleston News and Courier*, 4/27/64; *Washington Evening Star*, 4/29/64; *U.S. News and World Report*, 5/4/64; *Washington Post*, 4/4/64.

20. Billy James Harris, *Unmasking the Deceiver: Martin Luther King, Jr.* (Tulsa, undated).

21. *Congressional Record*, 3/30/65.

22. U.S. Senate, Committee on Commerce (88:1), *Civil Rights—Public Accommodations: Hearings on S. 1732* (Washington 1963), 375.

23. Ibid., 435–35.

24. Ibid., 359–63.

25. David J. Garrow, *The FBI and Martin Luther King, Jr.* (New York, 1981), 68.

26. Sullivan to Belmont, 8/30/63, Martin Luther King, Jr., FBI file (microfiche collection), (Frederick, Md., 1984).

27. Hoover testimony quoted in *Columbia Record*, 5/5/64.

28. Garrow, *The FBI and Martin Luther King*, 73.

29. Sullivan to Belmont, 12/24/63, King FBI file.

30. Garrow, *The FBI and Martin Luther King*, 125–26, 175, 183–86; Baumgardner to Sullivan, 1/9/64, King FBI file.

31. Memo, 3/4/68, in Perkus, *COINTELPRO*, 22–23; *Final Report*, 3:180.

32. Hoover in *New York Times*, 11/19/64.

33. Wallace quoted in Allen J. Matusow, *The Unraveling of America* (New York, 1984), 425.

Chapter 5   What Did We Know and When Did We Know It?: 1969–1977

1. Haldeman action paper, 6/15/71, in Bruce Oudes, ed., *From the President: Richard Nixon's Secret Files* (New York, 1989), 271–72.

2. Ehrlichman draft memo for Nixon, [6/15/71], ibid., 272–73.

3. Richard Nixon, *The Memoirs of Richard Nixon* (New York, 1978), 513.

4. Stanley Karnow, "Giap Remembers," *New York Times Magazine*, 6/24/90.

5. Nixon news conference, 7/25/69, *Public Papers*.

6. Arbatov in *Pravda*, 8/10/71, quoted in Raymond L Garthoff, *Detente and Confrontation* (Washington, 1985), 241–42.

7. *New York Times*, 7/22/71 and 7/28/71; *Time*, 7/2/71; *Newsweek*, 8/16/71; *Nation*, 9/6/71.

8. *Time*, 8/2/71; *New York Times*, 7/29/71.

9. "Basic Principles of Relations Between the United States of America and the Union of Soviet Socialist Republics," *Bulletin*, 6/26/72.

10. Henry Kissinger, *White House Years* (Boston, 1979), 638; CIA report quoted ibid., 639.

11. Kissinger news briefing, 9/25/70, quoted ibid., 646.

12. Ibid., 747.

13. Soviet note, 10/5/70, excerpted ibid., 649.

14. Nixon, *Memoirs*, 485.

15. Kissinger quoted in Philip J. Funigiello, *American-Soviet Trade in the Cold War* (Chapel Hill, N.C., 1988), 183; Nixon message to Congress, 5/3/73, *Public Papers*.

16. Lyndon Baines Johnson, *The Vantage Point* (New York, 1971), 305.

17. Kissinger, *White House Years*, 409.

18. Kissinger interview, 11/12/73, *Bulletin*, 12/10/73.

19. *Congressional Record*, 4/15/71, 7/27/71, 8/5/71, 9/8/71.

20. *Washington Post*, 7/30/71; *New York Times*, 8/1/71.

21. *Nhan Dan*, 8/17/72 and 8/19/72, quoted in Garthoff, *Detente and Confrontation*, 259.

22. *Congressional Record*, 12/15/75–12/19/75.

23. Ibid., 12/18/75, 12/19/75.

24. Unnamed FBI critic quoted in Stanley I. Kutler, *The Wars of Watergate* (New York, 1990), xii.

25. "The Enemy Within," *Progressive*, February 1975.

26. James Higgins, "Who Profits by Worldsnoop?," *Nation*, 4/5/75.

27. Roger Morris, "Banality of Power," *New Republic*, 10/4/75.

28. A. J. Langguth, "Abolish the CIA!," *Newsweek*, 4/7/75.

29. Jonathan Kapstein, "Philip Agee: The Spy Who Came In and Told," *Business Week*, 7/28/75.

30. James Burnhan, "Too Much Intelligence," *National Review*, 7/4/75.

31. Carter address, 5/22/77, *Public Papers*.

32. Congressional resolution quoted in Gaddis Smith, *Morality, Reason, and Power: American Diplomacy in the Carter Years* (New York, 1986), 50.

33. Carter to Sakharov quoted ibid., 67; Carter speech, 10/5/77, *Public Papers*.

34. Young speech, 12/14/78, in *American Foreign Policy: Basic Documents 1977–1980* (Washington, 1983), 432–34.

35. Vance speech, 7/1/77, ibid., 1131–35.

36. Carter address, 4/18/77, *Public Papers*.

Chapter 6   *Old Verities Die Hardest: 1977–1984*

1. Carter address, 7/15/79, *Public Papers*.

2. Peter J. Ognibene, *Scoop: The Life and Times of Henry M. Jackson* (New York, 1975), 154–55.

3. Ibid., 159–63.

4. Jackson statement, 12/22/70, and Jackson to Stennis, 12/21/70, in *Congressional Record*, 12/22/70.

5. Peter Steinfels, *The Neoconservatives* (New York, 1979), 29.

6. Theodore Draper, "Detente" and "Appeasement and Detente," *Commentary*, June 1974 and February 1976.

7. Norman Podhoretz, "Making the World Safe for Communism," *Commentary*, April 1976.

8. Strobe Talbott, *The Master of the Game: Paul Nitze and the Nuclear Peace* (New York, 1988), 106, 139–41.

9. Paul H. Nitze, "Assuring Strategic Stability in an Era of Detente," *Foreign Affairs*, January 1976.

10. Jerry W. Sanders, *Peddlers of Crisis: The Committee on the Present Danger and the Politics of Containment* (Boston, 1983), 201–2.

11. Jeane Kirkpatrick, "Dictatorships and Double Standards," *Commentary*, November 1979.

12. Jackson in Strobe Talbott, *Endgame: The Inside Story of SALT II* (New York, 1979), 5.

13. Carter interview, 12/31/79, in *American Foreign Policy: Basic Documents, 1977–1980* (Washington, 1983), 811–12.

14. Carter address, 1/23/80, *Public Papers*.

15. Brzezinski to Carter, 12/28/79, in Zbigniew Brzezinski, *Power and Principle: Memoirs of the National Security Adviser 1977–1981* (New York, 1983), 566.

16. Nixon toast, 2/24/72, quoted in Nixon, *Memoirs*, 580.

17. Carter speech, 6/7/78, *Public Papers*.

18. *Tass* quoted in *New York Times*, 6/8/78; *Washington Post*, 6/8/78; *New York Times*, 6/8/78; Oswald Johnston in *Los Angeles Times*, 6/11/78; McGrory in *Chicago Tribune*, 6/14/78.

19. Weinberger speech, 11/28/84, *American Foreign Policy: Current Documents 1984*, 65–70.

20. Shultz speech, 12/9/84, *Bulletin*, February 1984.

21. Reagan speeches, 2/24/82 and 3/23/83, *Public Papers*.

22. Reagan in *New York Times*, 10/26/83, and Reagan to Thurmond and O'Neill, 10/25/83, *Public Papers*.

23. Shultz interview, 1/22/84, *Bulletin*, April 1984; Eagleburger testimony, 2/2/84, U.S. House of Representatives (98:2), *The Crisis in Lebanon: U.S. Policy and Alternative Legislative Proposals: Hearings Before the Committee on Foreign Affairs* (Washington, 1984), 28–31; Reagan interview, 2/2/84, *Public Papers*; Reagan statement, 2/7/84, ibid.; Shultz testimony, 2/9/84, U.S. House of Representatives (98:2), *Foreign Assistance Legislation for Fiscal Year 1985: Hearings Before the Committee on Foreign Affairs* (Washington, 1984), 148.

24. Reagan address, 2/6/85, *Public Papers.*

25. *Statistical Abstracts of the United States 1990* (Washington, 1990), 378, 425, 428, 475.

26. Kevin Phillips, *The Politics of Rich and Poor* (New York, 1990), 83, 239.

27. *Statistical Abstracts 1990*, 306.

28. Standard & Poor's, *Analyst's Handbook 1989* (New York, 1989), 2–3.

29. *Statistical Abstracts 1989*, 327; *New York Times*, 7/22/85.

30. Reagan speech, 3/23/83, *Public Papers.*

31. Hans A. Bethe, Richard L. Garwin, Kurt Gottfried, and Henry W. Kendall, "Space-Based Ballistic-Missile Defense," *Scientific American*, October 1984.

32. Robert Jastrow, "The War Against Star Wars," *Commentary*, December 1984.

33. Matthew Rothschild and Keenen Peck, "Star Wars: The Final Solution," *Progressive*, July 1985; *New York Times*, 11/19/85.

34. *New York Times*, 7/22/85.

35. Rosy Nimroody, *Star Wars: The Economic Fallout* (Cambridge, Mass., 1988), 86; Gerald M. Steinberg, ed., *Lost in Space: The Domestic Politics of the Strategic Defense Initiative* (Lexington, Mass., 1988), 116.

36. Steinberg, *Lost in Space*, 116–19.

37. *Wall Street Journal*, 5/21/85; *Christian Science Monitor*, 9/17/84; *New York Times*, 11/19/85.

38. Steinberg, *Lost in Space*, 118.

39. Nimroody, *Star Wars*, 94–104, 112.

40. Warnke in Rothschild and Peck, "Star Wars"; Ball in Nimroody, *Star Wars*, 93.

41. *Statistical Abstracts 1989*, 303, 305.

42. Ibid., 303.

43. Norman Podhoretz, *Why We Were in Vietnam* (New York, 1982, 1983), 21, 48, 63, 103, 210.

44. Charles Krauthammer, "The New Isolationism," *New Republic*, 8/4/85.

45. George F. Will, column of 12/21/82, reprinted in *The Morning After* (New York, 1986).

46. Pro-Family Forum document, undated, quoted in Steve Bruce, *The Rise and Fall of the New Christian Right* (Oxford, England, 1988), 77.

47. Falwell quoted ibid., 150.

48. Richard A. Viguerie, *The New Right: We're Ready to Lead* (Falls Church, Va., 1981 ed.), introduction (unpaginated), 109–10, 119–22.

49. Hal Lindsey, with C. C. Carlson, *The Late Great Planet Earth* (New York, 1974 ed.), 41, 48, 54, 71, 85, 106, 154–56. Figure on book sales from Bruce, *Rise and Fall of the New Christian Right*, 87.

Chapter 7 Who Won the Cold War?: 1984–1991

1. Reagan speech, 1/16/84, *Public Papers.*

2. Bundy, Kennan, McNamara, and Smith, "Nuclear Weapons and the Atlantic Alliance," *Foreign Affairs*, Spring 1982.

3. Jonathan Schell, *The Fate of the Earth* (New York, 1982), 47–49, 183–84.

4. Krauthammer, "In Defense of Deterrence," *New Republic*, 4/28/82.

5. Ibid.

6. Shultz speech, 10/18/84, *Bulletin*, December 1984.

7. Weinberger to Reagan, 10/13/85, *New York Times*, 10/16/85.

8. *New York Times*, 10/17/85.

9. Reagan interview, 10/31/85, *Weekly Compilation*.

10. Reagan interview, 11/6/85, ibid.

11. Reagan interview, 11/12/85, ibid.

12. Talbott, *Master of the Game*, 285–86; Donald T. Regan, *For the Record* (San Diego, 1988), 317.

13. Talbott, *Master of the Game*, 311–12.

14. Reagan briefing, 8/6/86, *Weekly Compilation*.

15. Regan, *For the Record*, 337–55; Talbott, *Master of the Game*, 315–26; Ronald Reagan, *An American Life* (New York, 1990), 677.

16. William S. Cohen and George J. *Mitchell, Men of Zeal: A Candid Inside Story of the Iran-Contra Hearings* (New York, 1988), 123.

17. Perle in Talbott, *Master of the Game*, 346.

18. Rogers in *Washington Post*, 6/18/87; Shultz in *New York Times*, 6/21/87.

19. Reagan address, 12/10/88, *Weekly Compilation*.

20. George Kennan, "Obituary for the Cold War," *New Perspectives Quarterly*, Summer 1988.

21. Jeane Kirkpatrick, "The Withering Away of the State?" *New Perspectives Quarterly*, Winter 1988–89.

22. Krauthammer, "Beyond the Cold War," *New Republic*, 12/18/88.

23. Zbigniew Brzezinski, *The Grand Failure: The Birth and Death of Communism in the Twentieth Century* (New York, 1989), 1, 47–48, 242.

24. Francis Fukuyama, "The End of History?," *National Interest*, Summer 1989.

25. Dulles statement, 8/31/54, *Bulletin*, 9/13/54.

Index